Digital Methods in Developing Textile Products for People with Locomotor Disabilities

Bianca Aluculesei[1]

Sybille Krzywinski[2]

Antonela Curteza[1*]

Manuela Avadanei[1]

[1]The "Gheorghe Asachi" Technical University of Iasi, Romania
[2] Technische Universität Dresden, Germany

antonela.curteza@academic.tuiasi.ro

Published by **Materials Research Forum LLC**
Millersville, PA 17551, USA

Published as part of the book series
Materials Research Foundations
Volume 110 (2021)
ISSN 2471-8890 (Print)
ISSN 2471-8904 (Online)

Print ISBN 978-1-64490-154-0
ePDF ISBN 978-1-64490-155-7

This book contains information obtained from authentic and highly regarded sources. Reasonable efforts have been made to publish reliable data and information, but the authors and publisher cannot assume responsibility for the validity of all materials or the consequences of their use. The authors and publishers have attempted to trace the copyright holders of all material reproduced in this publication and apologize to copyright holders if permission to publish in this form has not been obtained. If any copyright material has not been acknowledged, please write and let us know so we may rectify in any future reprint.

Distributed worldwide by

Materials Research Forum LLC
105 Springdale Lane
Millersville, PA 17551
USA
http://www.mrforum.com

Printed in the United States of America
10 9 8 7 6 5 4 3 2 1

Table of Contents

Introduction

Disability can be defined as the incapacity, or the diminution of the ability, to perform certain activities, or to exhibit behaviors, which are normally connected with everyday life. A disability can result from a specific impairment originating from a range of physical, cognitive, intellectual, sensory or mental issues. Over the years, the term 'disability' has come to be understood as a complex experience; starting with the medical aspects of what the disability means, but also its implications for the problems at the social level which then impact the individual.

Disabled people represent 15% of the total world population and, within the EU, disability afflicts some 10% of the overall population; with a significant probability of the numbers increasing in the near future [1,2]. This category or group of people constitutes a high concomitant demand for the marketing of personalized and functional products.

Disabilities impose special functional requirements with regard to clothing products, and increase the necessity for physical and psychological comfort. The wearer is typically searching for clothing products having divers functional parameters (e.g., protection, comfort, ease of movement, fit) as well as the fashionable aspects needed to guarantee social ease (e.g., self-esteem, group-acceptance, respect of others). Those suffering from disabilities have to consider, when buying clothes, their different body-shape, limited mobility, medical problems and psychological and social needs.

The clothing industry has an insufficient choice of products on offer, with which to meet the needs and demands of disabled people, because garments intended for the latter impose specific design requirements. Most of the garments for disabled people which are currently on the market are not made of fabrics that are adapted to their medical problems. The products also do not have a pattern construction which is suitable for accommodating atypical bodies, postures or non-typical movements. More attention needs to be paid to the research and development of functional clothing for disabled people so as to improve functionality, attractiveness, ease-of-use and affordability.

In order to produce a customized product that can fulfill the necessities of a target market, it is first necessary to conduct intensive research into the characteristics of that group, especially when it includes people possessing a variety of physical abilities. Market research in this field can be more complex and challenging because of the variety of needs and characteristics of disabled persons. The clothing needs of those living with some sort of disability are not being met in general: there is an absence of appropriate clothing, which hinders this group of people from pursuing normal social activities and relationships, jobs or just everyday life [3,4].

The variety of disabilities which lead to special design requirements for clothing is vast, and each one needs to be studied carefully. This work analyzes some aspects which govern the needs of persons with paraplegia who use a wheelchair for mobility.

Paraplegia refers to an impairment, or loss, of motor and/or sensory functions in part of the spinal cord. It can affect the functions of the trunk, legs and pelvic organs [5]. Wheelchair-users make up 1% of the European population; nearly eight million people [6].

Wheelchair-users are very sensitive, to the clothes that they wear, with regard to their functional and design characteristics. They have to consider their health problems (e.g., skin fragility) and the fit of the clothes to their body; which is problematic due to their body-shape and sedentary life. The main desires of the wheelchair-user are functionality, attractiveness, ease-of-use, affordability and safety.

It is to be concluded the designing of clothes for wheelchair-users imposes special requirements that result from their body-shape and sedentary position. The fabrics used must also possess good mechanical characteristics and thermal properties, so as to ensure the comfort of the wearer.

Chapter 1

1. Disabilities: Classifications and Clothing Needs

1.1 Disability

1.1.1 Definitions and classifications

Disability is defined by the *International Classification of Impairments, Disabilities, and Handicaps* (Geneva 1980), in the context of health experience, as a limited participation in daily living activities that comes from a physical or psychological impairment [7]. At another level of understanding, with the development of the *International Classification of Functioning, Disability, and Health* (Geneva 2001), the World Health Organization offers a broader meaning of *disability*, that covers impairment, activity limitation and restricted participation [8]; thus bringing together aspects of both health and social problems. The medical aspect of disability refers directly to the health problems of an individual which various diseases can cause, trauma or other causes. The social aspects simultaneously take disability to be part of a deeper malaise which touches on the difficulty of integrating such individuals into society. It is clear that disability encompasses not just a medical feature but also a complex set of environmental factors [9].

Several types of disability classification have been drawn up, the most widely used being *The International Classification of Functioning, Disability, and Health,* recommended in 2001 by the World Health Organization. The *International Classification of Impairments, Disabilities, and Handicaps* (ICIDH), published in 1980, is the predecessor of the *International Classification of Functioning, Disability, and Health* (ICF) and describes a more common framework and the definitions of disability and its existing problems.

a. International Classification of Impairments, Disabilities, and Handicaps (ICIDH)

The ICIDH highlights three aspects to be studied in order to understand the situation of people with a disability: *impairment* - abnormalities of a body-structure, organ or system function (Table 1.1.), *disability* - consequences of the impairment in terms of functional performance and activity (Table 1.2.) and *handicap* - disadvantages experienced by the individual as a result of impairments and disabilities (Table 1.3.). According to this classification, disability represents an impairment, and it reflects a particular morphology of the body and disturbances of the person on a social level [7].

Table 1.1. Classification of impairments according to ICIDH [7]

Type of impairment	Affected areas
Intellectual	Intelligence, memory and thought-processes.
Other psychological impairments	Presence of neurophysiological and psychological diseases.
Language	Comprehension and use of language and its associated functions, including learning.
Aural	The hearing function and its associated structures.
Ocular	The vision function and its associated structures.
Visceral	The internal organs and other special functions.
Skeletal	Mechanical and motor disturbances of the face, head, neck, trunk and limbs, as well as deficiencies of the limbs.
Disfigurement	Impairment having the specific potential to interfere with or disturb social and interpersonal relationships.
Other impairments	Including multiple impairments, severe incontinence, metabolic impairment, other generalized impairment, sensory impairment of head, trunk, upper limbs or other body-part.

Table 1.2. Classification of disabilities according to ICIDH [7]

Type of disability	Affected areas
Behavior	The awareness of the individual and his ability to accomplish everyday activities and to interact with others, including the capacity to learn.
Communication	The ability of the individual to create and transmit messages and to receive and understand messages.
Personal care	The capability of the individual to care for her/himself with regard to basic physiological activities, such as feeding, excretion, hygiene and dressing.
Locomotor	The ability of the individual to perform activities such as moving himself, or objects, from one place to another.
Body disposition	The ability of the individual to complete various actions related to the moving of parts of the body, including secondary activities such as executing tasks associated with her or his domicile.
Dexterity	Adroitness and skill in bodily movement, including manipulation and the control of mechanisms.
Situational disabilities	Disabilities related to dependences and endurance, environment or other situational disabilities.
Particular skill	Behavioral and task fulfillment abilities.
Other activity restrictions	

A *disability* is characterized by the loss of the ability to perform certain activities or behaviors connected with essential everyday living actions; for example, in personal care (the ability to wash or feed oneself), in other daily activities (such as communicating with other people) or locomotor activity [7].

Table 1.3. Classification of handicaps according to ICIDH [7]

Type of handicap	Affected areas
Orientation	The ability of the individual to orient himself in relation to his surroundings.
Physical independence	The ability of the individual to maintain an effective and independent daily existence.
Mobility	The ability of the individual to move effectively within his surroundings.
Occupation	The ability of the individual to pass the time in a manner normal for his/her gender, age and culture.
Social integration	The ability of the individual to establish, participate in, and maintain social relationships.
Economic self-sufficiency	The ability of the individual to sustain daily socio-economic activity and independence.
Other handicaps	Additional factors that may give rise to disadvantage.

b. International Classification of Functioning, Disability and Health (ICF)

ICF focuses on the importance of environmental factors that can impair or improve the life of a disabled person. This is the main difference between the new classification and the previous one, ICIDH [10], which was developed in order to describe the health status of populations on a global scale. ICIDH was more focused on disease and was unsuccessful in capturing the effect, upon adequate functioning, of the social and physical environment [11]. ICF classifies disability groups into two parts, having differing components for a person with a specific health condition [12], firstly the Components of Functioning and Disability and, secondly, any Contextual Factors [9]. ICF uses numerical codes that specify the amplitude or dimensions of functioning and disability, within a specific category of environmental factors, that could be a facilitator of, or barrier to, good performance [13].

The Components of Functioning and Disability refer to the limitations on activity and restrictions on participation that an individual may suffer due to various impairments arising from the loss or malfunction of a bodily function or structure. The Contextual Factors reflect the background of a person's nature and lifestyle. They can be divided into two parts: environmental and personal. The environmental factors bring together the physical or social environment within which people pass their lives, e.g., home, school, workplace or any manner of cultural system. The personal factors take account of individual features such as gender, race, sex-life, social background, education etc. [9].

ICF aims to combine aspects of both the medical and the social model, and it conceptualizes disability as being a reciprocal interaction between health problems (diseases, disorders, injuries, traumas, etc.) and contextual factors. Each of these components can be found within numerous domains and, within those domains, numerous categories can be found which constitute the classification of ICF [9].

1.1.2 Physical disabilities

Upon analyzing the complex definitions and classifications that exist with regard to disability, it can be simply concluded that a disability can occur due to physical, mental or sensory impairment [14]. Every impairment that leads to a disability can be either congenital, or acquired due to aging or traumatic accident.

A physical disability can be defined as being one which reduces a person's mobility, sight, hearing, powers of communication and/or the ability to perform everyday activities [15]. The functional abilities that are typically affected by a mobility disability are: walking, climbing stairs, sitting-down or standing-up, taking care of various chores, maintaining balance and coordination, lifting, stretching, as well as performing routine daily activities [16].

Two groups of conditions having a physical origin affect mobility: muscular/skeletal and neurological [17]. A muscular/skeletal condition implies a malformation of the skeletal system, and the impairment can involve a congenital anomaly, an illness or an accidental trauma. Neurological conditions that implicate the nervous system affect the movement, use or control of certain parts of the body. These impairments can again result from a congenital anomaly, a disease or an accident (Table 1.4.)[18].

Table 1.4. The Three Classifications of Disability [18]*

Named Condition	Description	Physical manifestation/s
Classification 1: Congenital origins stemming from hereditary genetic factors (birth defects)		
Spina bifida (p.151)	Abnormal intrauterine development of the lower portion of the spinal cord	Partial or complete lower extremity paralysis and an inability to develop bowel and bladder control
Down syndrome (p.151)	Chromosome abnormality manifesting in at least 50 clinical signs	Wide variations in mental ability, behavior, and developmental progress

Digital Methods in Developing Textile Products for People with Locomotor Disabilities
Materials Research Foundations **110** (2021)

Materials Research Forum LLC
https://doi.org/10.21741/9781644901557

Classification 2: Developmental origins diagnosed from specific symptoms that increase in severity		
Cerebral palsy (p.152)	Most common developmental disability affecting the musculoskeletal system. stems from malfunction of the central nervous system due to brain damage	Developmental problems including abnormal motor control, limited mobility and self-care skills
Stroke (p.703)	Acute onset of nervous system dysfunction. Largest single cause of neurologically based disability in the US.	Weakness or paralysis on one side of the body; problems with balance; disturbance in sensation and disruption of speech
Multiple sclerosis (p.505)	Disease of the central nervous system white matter	Motor impairment including difficulty walking, and increased fatigue
Muscular dystrophy (p.506)	An inheritable chromosome-based disease with many different forms	Progressive muscular weakness and atrophy
Musculoskeletal disorders (p.478)	Broad spectrum of conditions that affect multiple soft tissues and bone	Alterations in structural integrity resulting in pain, altered function and disability
Polio (p.512)	Neuromuscular disorder; virus which causes deterioration of motor neurons	Respiratory failure; possible paralysis; atrophy in groups of muscles, contraction and possible deformity
Rheumatic disorders (p.479)	Degeneration of joints through inflammation	Impaired mobility in joints, chronic pain, stiffness, fatigue, physical appearance changes
Scoliosis (p.490)	Structural bone disorder; lateral deviation in the normally straight vertical line of the spine	Deviations in appearance – S-curve in spine, sagging shoulders, noticeable asymmetry to upper part of body in one or more areas
Classification 3: Acquired disability, occurring through accident, disease or injury		
Partial or complete paralysis	**Paraplegia:** paralysis of the lower limbs, part or whole of trunk - usually a result of injury to back	Loss of use and sensation in lower body and part of trunk
	Quadriplegia: paralysis of all four limbs and trunk - usually a result of injury to neck	Loss of use and sensation in all four limbs
Amputation	Loss of one or more limbs (majority of upper limb amputation through accident; majority of lower limb amputation through vascular deficiency)	Loss of mobility; changes in appearance

*Note. Definitions obtained from Encyclopedia of Disability and Rehabilitation by A. E. Dell Orto, and R. P. Marinelli, (1995).

Three major origins of disability exist: congenital, developmental and acquired. Overlaps exist; for example, amputation can be caused by a developmental disease or an accident. The table shows each of the three and *some* examples within each group with characteristics and physical impact (from Dell Orto and Marinelli, 1995).

1.1.3   Acquired disabilities - Spinal cord injury

A *spinal cord injury* (SCI) means an injury to the spinal cord or nerves which can affect the motor and/or sensory abilities of the person and his other bodily functions [19]. Spinal cord injuries can be classified into two main groups: traumatic and non-traumatic. A traumatic SCI is the consequence of contusion, compression or excessive stretching of the spinal cord. A non-traumatic SCI results mainly from spondylosis, vascular ischemia, tumor compression or congenital and inflammatory spinal cord disorders [20].

The physical impairments caused by SCI may differ, depending upon the position and severity of the injury. The occurrence of a SCI can affect almost every aspect of the quality-of-life of a person, such as physical health, employment, personal relationships and cherished hobbies.

The disabilities that stem from spinal cord injuries can include impairment or loss in the areas of motor activity, sensory functions, incontinence, reflexogenic movements and muscular hypertonicity (spasticity), flaccid paresis (lower motor neuron syndrome), impairment of autonomic functions (vasomotor or sensorimotor control) and metabolic or hormonal imbalances [21]. In addition to the disruption of everyday life, the lack of any specific treatment for SCI is another factor which impacts the life of a person suffering from SCI. Existing alternatives for reducing the suffering of SCI patients are, up until now, largely palliative [22]. The level of a lesion that is implicated in a spinal cord injury is defined by its position in the human spine. Thirty spinal cord segments can be identified: 7 being cervical, 12 being thoracic, 5 being lumbar, 5 being sacral and the thirtieth being the coccyx region (Fig. 1.1) [23].

Damage to the upper part of the spine generally causes more significant deficiencies than does damage to the lower part. The cervical area which is affected causes leg and trunk paralysis, plus sensory loss. Damage may also cause partial paralysis of the arms, unlike damage to the thoracic and lumbar spine, leading to paralysis of the legs alone [24]. Alteration of the various spinal cord segments produces the following divers effects [25]:

- C1 - C4: breathing and, C2, head and neck movements,
- C4 - C6: heart rate and, C5, shoulder movements,
- C6 - C7: wrist and elbow movements,
- C7 - T1: hand and finger movements,
- T1 - T12: sympathetic tone and, T2 – T12, trunk stability,

- T11 - L2: ejaculation and, L2, hip movement,
- L3: knee extension,
- L4 - S1: foot motion and, L5, knee flexion, and
- S2 - S4: S2 – S3 bowel and bladder activity.

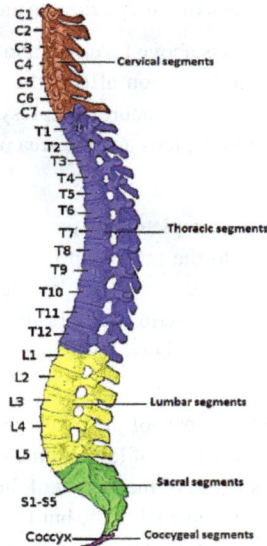

Figure 1.1. Spine segments (Wikipedia source: Henry Gray (1918) - Anatomy of the Human Body)

1.1.4   Types of paraplegia - body morphology and medical problems

Spinal-cord lesions which arise due to injury or disease can produce tetraplegia or paraplegia, depending upon the level of its location. The severity of the injuries can be broadly classified into two types [26]: complete - no sensory or motor function below the level of the SCI (specifically at S4–S5) or incomplete – some sensory and muscular functions are present below the neurological level of the injury, including the lowest sacral segments (S4–S5) [27]. Complete injury is defined as being a total loss of sensory and motor functions resulting from damage to the lowest sacral area [21], while incomplete injury is defined as being a partial loss of sensory and/or motor functions below the current level of the injury [21].

Paraplegia refers to an impairment or loss of motor and/or sensory functions in the thoracic (T2-T12), lumbar (L1-L5) or sacral (S1-S5) segments of the spinal cord [5]. It can affect the functions of the trunk, legs and some of the pelvic organs, but the upper part of the body and the arms are unaffected. It can be caused by damage to the spinal cord during an accident, damage due to the occurrence of tumors, tuberculosis or transverse myelitis, or can be hereditary. There are two types of paraplegia: spastic and flaccid [28].

In spastic paraplegia the spine lesion is above level T1, and the result is a loss of sensation from that point downwards. When the lesion affects the spinal cord below level L1, it involves the cauda equine [29] and the neurological symptoms are those of flaccid paraplegia; involving sensory loss in the legs and the area of the perineum [30].

a. Complete injury in paraplegia

Complete paraplegia describes the complete and permanent loss of the ability to send sensory and motor nerve impulses to the muscle groups which are controlled by nerves extending from the spine to the T1 level or below. People with T1 lesions can use their arms, and they retain all of the motor functions of a healthy person. The sensory function is lost at the level of the lesion, and the bowel and reproductive organs no longer work properly.

With complete paraplegia in the lower part of the body, some patients can partially move the trunk and can stand up using some form of long orthosis (which supports the paralyzed legs), and a frame which permits support of the weight of the body by using the arms. They can move over short distances by using such aids, but they must be supervised. Patients with T6-T12 lesions also have partial muscular control in the abdominal area, and they can move on their own over short distances with the aid of orthotics, a frame or carriage which supports the weight of the body with the help of the arms. The most common mobility solution for those who are completely paralyzed is the wheelchair [31].

b. Incomplete injury in paraplegia

Those who suffer incomplete paralysis have a different level of disability, and each one is unique. Trauma to the chest and below can lead to paraplegia without affecting the arms. In the case of lesions between levels T1 to T8, these often present as control of the arms but weak control of the trunk, due to the lack of control of the abdominal muscles, and balance in the sitting position is generally reasonable. Lumbar and sacral trauma leads to decreased control of the hip and lower limb flexors.

Other changes in bodily functions can occur, such as poor bowel and bladder control. Among the main problems are: low blood-pressure, reduced control of body temperature, inability to sweat in the inferior area of the lesion and chronic pain [31].

The expectation of recovery of persons suffering from complete paraplegia is 5% and, the lower the level of the injury, the greater is the possibility of healing. Those afflicted with incomplete paraplegia enjoy the best prognosis with regard to ambulatory recovery. Among the total number of persons initially suffering from incomplete paraplegia, 80% recovered control of their hip flexors and knee extensors within one year [26].

c. Flaccid paraplegia

Flaccid paraplegia is characterized by a rapid onset of weakness in both legs and by a weakness of the muscles of the respiratory and deglutitional systems; progressing to maximum severity within several days or weeks [32]. A clinical description of flaccidity would be reduced muscle tone, with the muscles losing their ability to contract and become limp [33]. Absence of the knee-jerk reaction may also be observed, with a negative Babinski sign and no meningeal irritation [34]. Other characteristics of flaccid paraplegia are sensitivity disorders, urine-retention plus even urinary tract infections and oedemas which can eventually lead on to bed-sores and even ulcers [31].

d. Spastic paraplegia

Hereditary Spastic Paraplegia (HSP), also known as SPG (Spastic Paraplegia), is an inherited neurological disease [35]. As well as its primary clinical manifestation as a weakness due to degeneration of the orticospinal tract, spastic paraplegia also comes with slow, stiff and jerky knees and dragging of the toes. The gait is cumbersome and spastic, with abduction of the thighs. The patient can crawl using the legs in a club-foot position and adopt a jump-like walk, with the feet leading, with the support of a carriage [31].

1.2   Functional clothing for persons with special needs

The field of functional clothing can address various needs, from the protective to the aesthetic. It includes specially designed clothing which exhibit specific functional characteristics [36].

Most of the garments for disabled people which are currently available on the market suffer from problems that are caused firstly by the characteristics of the fabric (the properties of the latter not adapted to the medical problems, mechanical or physiological, of the individual) and, secondly, by unsuitable pattern designs for handling the atypical body-shape, posture or movements of the wearer [37]. Much greater attention must therefore be paid to the research and development of clothing for special needs, so that a person suffering from a disability can enjoy more suitable personalized clothing.

Garments for persons with special needs are part of the functional clothing category, and are meant to improve the quality of life of those whose body-shape, mobility or dexterity differs from the norm. Paraplegic persons are a part of this category of people, martyrs to their clothes, given the current functional and design characteristics of the latter.

From the psychological point of view, a person with a disability faces the everyday burden of being different and of having special needs. A disabled person, just like a healthy one, also has definite expectations with regard to clothing. He will be interested in clothing from different points of view, such as functionality, comfort or durability and fashion. Proper clothing can help disabled people with their rehabilitation and integration into society. Clothing should in general satisfy various functional requirements (climatic and environmental protection, ease of movement, lightness, fit, limited pressure on the body, low friction against the body, tactility, lack of static charging) and symbolic aspects which address social requirements (self-esteem, state of mind, group acceptance, decoration, fashion, respectability). Those with disabilities are also faced with changing body-shape, mobility limitation, medical problems and special psychological and social support [36,38]. Clothing for disabled people must, in other words, satisfy the requirement [16,39] to:

- match the disability and keep the handicap under control by adequate pattern design - including fastenings, posture matching, allowance for joint mobility;
- ensure independence of movement and use of fasteners, plus a high level of comfort and safety, ease of use and wear/tear-resistance;
- allow easy maintenance (ease of washing and ironing plus stain-repellence/resistance);
- be made from fabrics that will not chafe nor irritate the skin, and ensure good thermal isolation plus moisture-management and perspiration-control;
- be made from natural fibers, or treated with antibacterial solvents, as well as satisfying anti-microbial/bacterial requirements;
- provide a certain psychological comfort, and impart self-confidence to the wearer.

1.3   Needs and demands of wheelchair-users with regard to clothing

Wheelchair users represent, according to the World Health Organization, about 15% of the worldwide population which lives with a form of disability [1]. Disabled people in the EU represent about 10% of the total population, and it is estimated that this percentage will increase to 29% by 2050 due to aging and disease. Because of their permanent or semi-permanent sedentary position, the associated contact area must support the weight of their body for long periods. The absence of muscle tone in the legs, and the lack of activity, reduces the normal cushioning effect. It decreases the circumference of the legs, while possibly increasing that of the upper part of the body, and renders the circulatory system less efficient. The nerves of the spine being damaged, there is an absence of sensory feedback, thus making the dermal tissue especially vulnerable to oedema and ulceration. Because of the prolonged time spent in the sitting position, perhaps combined with incontinence, the moisture-exposure of the affected areas is high and makes the skin even more sensitive.

The main attributes of clothing, such as functionality, attractiveness, ease-of-use, affordability and safety [40], are essential to disabled people. The requirements with regard to clothing for wheelchair-users must make even greater aesthetic, functional and accessibility demands [41] because of their particular health problems (skin fragility and ulceration, venous diseases, hygiene difficulties, high-sweat areas).

1.3.1 Aesthetic demands

Clothing products are essential, at the aesthetic level, for a good self-image. This is especially so for a person who has a disability and for whom a garment can increase morale and provide psychological comfort. If his clothes are aesthetically pleasing, the disabled person can feel safer, more confident and feel able to integrate more successfully into society.

The designing of clothes for wheelchair-users must confront some particular problems involving the shape of the body in the sitting position, which has to be considered when creating patterns. The practical aspects of the clothing, and the aesthetic problems which can arise, have to be analyzed in greater detail due to the tendency of the fabric to bunch-up in certain areas (e.g., stomach, back of the knee), to expose the back because of inadequate fitting of the garment to the buttocks or to leave the legs uncovered due to an improper choice of garment length.

From the aesthetic point of view, another problem is the type of fabric from which the clothes are to be made. There is a need for fabrics which possess good mechanical characteristics at those areas of the product where abrasion can be considerable (e.g., buttocks, back or elbows). The thermal properties of the fabric are also important because thermal imbalance and sweating can also cause aesthetic problems [41,42].

1.3.2 Functional demands

Several functional characteristics should be taken into account when creating clothes for wheelchair-users. Comfort is an essential require

ement in the case of clothing for special needs. Textile products that come into contact with the body should not cause any discomfort to the wearer. Various aspects of comfort that should be addressed are thermal insulation and permeability, tactile factors, usability, fit and treatment with hypo-allergenic agents [16,6,42].

a. Fitting and freedom-of-movement problems:

- the body posture and position influence the fitting of clothing to the body and freedom of movement in the wheelchair,

- the clothing must have a volume which fits into the space available in the wheelchair,

- the product should permit widening in the area of the neck, chest, abdomen, arms, hips and legs in order to provide better freedom of movements,

- the product should permit extension of the length of pants, and of the backs of shirts, jackets and coats, and permit the reduction of accumulated excess fabric at the waist or at the back of the knees.

b. Usability aspects:

- the need for special positioning of the closure system in order to facilitate dressing/undressing,

- the need for closure systems (zippers, Velcro® strips, buttons, snaps) which are adequate for handling the clothes,

- the need for easy-fastening systems which facilitate lavatory visits, plus openings for the passage of catheters or urine bags,

- the need for durable and resistant fabrics which are easy to clean, and

- the need to increase the resistance of materials in high-friction areas such as the buttocks and elbows, in order to ensure longer life for the clothes, while simultaneously avoiding thick and hard seams, especially in areas such as the back and buttocks which are exposed to high levels of pressure, that might lead to pressure-sores and wounds.

c. Thermal insulation and permeability aspects:

- the existing moisture in the garment or between the garment and the chair may contribute to skin problems such as ulcers or pressure-sores, leading to a need for absorbent fabrics that can ensure moisture management in areas having a higher level of sweat, such as the buttocks, chest, dorsal, lumbar, neck and abdominal regions,

- the wheelchair-users suffer thermal discomfort in the lower limbs and hands, and so the fabric should assure a thermal insulation which is adequate to protect against both cold and hot weather, and

- the thermal comfort depends upon the layer between the skin and the environment with respect to how well it permits the evaporation of sweat and the transmission of heat so as to maintain an equality between heat production and heat loss and thus a thermal balance.

d. Tactile factors:

- tactile comfort is essential to wheelchair-users because of the prolonged periods spent in a sedentary position, leading to a need for fabrics which exhibit mechanical properties such as high surface smoothness and elasticity,

- the need for elastic fibers which ensure superior comfort, and

- the clothing must generate minimal levels of electrostatic charge because the latter can cause clothing to stick to the body or attract dust, leading to a lower permeability or hygiene problems.

e. Hypo-allergenic agents:

- the need for anti-bacterial treatment of the fabric, or the use of a combination of natural and synthetic fibers, in order to ensure minimal body-odor retention or maintain good fabric permeability.

The current garment marketplace for people with disabilities is not well-developed; consumers with special needs often have problems in finding garments which are customized to their requirements. This is because of the number of product adaptations which would have to be performed, such as adding particular elements, making individual patterns, ensuring exceptional mechanical properties in the materials used, employing special yarns or carrying out unique functional treatments. Table 1.5 is a list of some of the producers and distributors of garments for disabled people.

Table 1.5: List of producers and distributors of clothing for disabled people

Name	Web address
LegaWear, Sweden	http://legawear.com
IZ Adaptive Clothing, Canada	http://www.izadaptive.com
AG Apparel, USA	http://www.agapparel.com
Clothing Solutions for Disabled People, England	http://www.clothingsolutions.org.uk/
Able2Wear, Scotland	http://www.able2wear.co.uk/
Buck & Buck, USA	http://www.buckandbuck.com/
Silvert's Adaptive Clothing & Footwear, Canada	https://www.silverts.com
RULATEX GmbH, Germany	https://www.renato.de/
Rollitex England	www.rollitex.co.uk
Rollimoden, Germany	https://www.rollimoden.de/

Chapter 2

2. Garment Design Methods for Wheelchair-Users

2.1 Anthropometric measurements

The product design of any garment must be based upon information concerning the body-shape and dimensions of the wearer. *Anthropometry* is the discipline of measuring the dimensions of the human body [43]. The dimensional characterization of the latter involves the listing of a series of sizes, parameters and anthropometric indicators [44]:

➢   *dimensions,*

- linear dimensions: height, diameter and depth.

- curvilinear dimensions (measured following the surface of the body): length, width and circumference.

➢   *angles,* and

➢   *weight* (kg).

In any anthropometric study, it is essential to follow the methodology prescribed by established protocols. Manual measurement research [43] specifies the need to mark the position of the anthropometric point. The latter is an easily identifiable reference-mark placed on the surface of the body, as determined by the bone-structure, by a well-defined boundary between soft tissues or by an epidermal mark on the body of the subject to be measured. It serves as a fixed point for the sampling of various body dimensions. Anthropometric studies typically use about 100 anthropometric points, but garment construction requires only 20 to 27 points [44]. The body dimensions which are usually measured for the purposes of garment construction are listed in Table 2.1.

Table 2.1: Body dimensions required for garment construction [45]:

Dimension	Definition
Height	The vertical distance between the highest detectable head point and the ground, with upright body and closed legs.
Bust/Chest girth	The maximum horizontal circumference measured under the armpits, and the level of the maximum projection of the bust/chest during normal breathing.
Under-bust girth	The horizontal circumference of the body, immediately below the breasts.
Waist girth	The horizontal circumference at the natural waistline between the highest part of the hip bones and the lower ribs.
Hip girth	The maximum horizontal circumference of the trunk, measured at hip height and at the fullest part of the buttocks.
Outside leg length	The distance from waist level to the ground, first following the contour of the hip, and then going vertically downward.
Waist height	The vertical distance from waist level to the ground.
Inside leg length	The vertical distance between the crotch level (mid-sagittal plane), with legs slightly apart; the weight being placed equally on each leg.
Neck girth	The circumference of the neck column, measured close to the larynx cartilage prominence and the 7th cervical vertebra.
Neck-base girth	The circumference around the base of the neck column, measured from the 7th cervical vertebra over the neck shoulder points and the center front neck point and back to the nape point.
Neck base diameter	The horizontal distance at the neck base between the left and right shoulder neck points.
Shoulder length	The direct distance from the base point of the neck to the shoulder point (acromion extremity), with arms hanging.
Shoulder slope	The inclination angle of a straight line joining the shoulder neck point and the acromion to the horizontal.
Back width	The horizontal distance from armpit level to armpit level, measured on the back.
Trunk length	The vertical distance from the level of the 7th cervical vertebra to the crotch level.
Back waist length	The vertical distance from the 7th cervical vertebra to the waist, measured along the contour of the spinal column to the waist level.

Dimension	Definition
Cervical height	The vertical distance from the 7th cervical vertebra, first following the contour of the center of the back to the hip level and then straight down to the ground.
Cervical to knee hollow	The vertical distance from the 7th cervical vertebra, first following the contour of the center of the back to the hip level and then straight to the hollow of the knee.
Neck point to breast point	The length from the base of the neck point to the breast point.
Bust width	The horizontal distance between the bust points (nipples).
Bust/chest height	The vertical distance between the level of the maximum bust/chest projection and the ground.
Front waist length	The distance along the contour from the neck base point to the nipple, and then vertically straight down to the front waistline.
Waist to hip	The distance along the contour of the side of the body, from waist level to hip level.
Total crotch length	The distance from the center of the waist level at the front of the body, through the crotch to the center of the rear waist level, following the contours of the body.
Natural waist circumference	The horizontal circumference at the natural waistline, between the highest point of the hip bones and the lower ribs.
Arm length, bent	The distance from the armscye/shoulder line intersection (acromion) to the prominent wrist bone, measured over the elbow (with the arm bent at 90°).
Upper arm length	The distance from the armscye/shoulder line intersection (acromion) to the outside elbow joint (olecranon), with the arm bent at 90°.
Upper arm girth	The maximum circumference of the upper arm and the lowest scye level at the arm base, measured with arms hanging.
Wrist girth	The girth of the wrist measured over the wrist (carpal) bones.
Elbow girth	The circumference of the arm at the elbow, with the arm bent at 90°.
Thigh girth	The horizontal circumference at the highest thigh position.
Knee girth	The circumference of the knee at the tibia level.
Knee height	The vertical distance from the tibia level to the ground.
Calf girth	The maximum girth of the calf, measured horizontally at the fullest part of the calf with legs slightly apart.

Dimension	Definition
Minimum leg girth	The minimum girth of the lower leg, measured horizontally just above the ankle.
Middle hip	The horizontal circumference of the trunk, measured midway between waist level and hip level.
Arm length – cervical point-acromion-elbow-wrist	The distance from the cervical point (7th vertebra), over the top of the shoulder, along the arm (bent horizontally at 90°) to the wrist bone.
Head circumference	The maximum horizontal girth of the head, above the ears.
Shin bone (cnemis) height	The vertical distance from the ground to the tibia level.
Upper knee girth	The circumference of the leg just above the kneecap.
Crotch length front	The distance along the contour, from the center of the waist level at the front of the body, to the crotch.
Middle forearm girth	The girth of the forearm at the middle between, elbow and wrist, orthogonal to the axis of the forearm.

There are two methods for performing human body measurements: direct (contact) and indirect (non-contact) [44,46,47]:

> *contact methods:* traditional measuring tools,
> *non-contact methods*: scanning and photogrammetry.

2.1.1 Traditional measuring methods

In the traditional or direct contact method, human body dimensions are measured between anthropometric reference-marks while following a specific protocol (according to the number of standard anthropometric measurements). This method is less costly, and has the advantage of being able to make direct observations of the postural peculiarities, conformation and physiognomy of the subjects measured. It involves the use of a standard position, and measuring the distances between body reference-marks by using stadiometers, anthropometers, sitting-height tables and tape-measures [44,48]:

> anthropometers – for measuring heights,
> tape-measure, roulette – for making curvilinear measurements,
> anthropometric compass – for measuring diameters,
> triangle, ruler – for measuring depths, coordinates,
> balance – for measuring weight.

The main disadvantages of this method are:

> ➢ the necessary immobilization of the measured person for a long period and the appearance of errors in the measurements due to inaccurate use of the instruments or to the incorrect choice of anthropometric points,
> ➢ the high volume of work involved in performing anthropometric research and obtaining data, without having a digitized 3D view of the measured body, and
> ➢ the low reproducibility of the data under the same measurement conditions.

2.1.2 Scanning methods

A 3D scanner is a tri-dimensional measurement device which is used to capture the shape of real objects or environments, to be then modified and analyzed in digital format. It can be used to obtain full or partial 3D data on the scanned object or environment. The result of scanning is a polygonal object which can be rotated on a computer screen so as to observe the shape from various angles. 3D scanners are now being used for a wide spectrum of applications, such as industrial engineering (e.g., automotive products), medical, sports performance, garment creation, animation, computer games, anthropometry, ergonomics, art, architecture etc. [49,50,51]. A 3D scanner consists of [52]:

- a light-source/laser that is projected onto the object/body,
- cameras that capture images of the lights projected onto the object/body,
- software which follows a standard procedure for measuring the object/body dimensions, and
- a computer for visualizing the scanning procedure and the resultant 3D object, and listing (as codes, values, etc.) the measured dimensions.

Figure 2.1. Desktop Scanner – EinScanSE Shining 3D, Source SHINING 3D Technology [53]

Figure 2.2. Handheld Scanner - Eva, Artec, Source Artec 3D [54]

Figure 2.3. Industrial Scanner - <u>Space Spider</u>, Artec, Source Artec 3D [55]

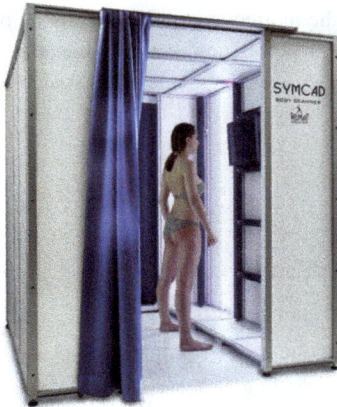

a) Body scanner SYMCAD III (standing position)	*b) Body scanner SYMCAD II (sitting position – can be used for people with locomotor disabilities measurement's, with a special application called STT)*

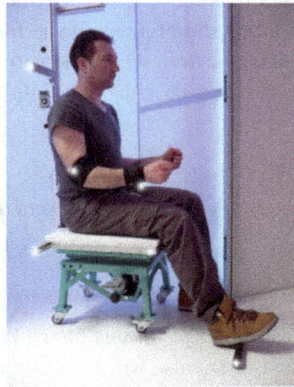

Figure 2.4 Body Scanners -TELMAT Industrie, Source TELMAT Industrie [56]

On the basis of their designs, technical designations and fields of application, 3D scanners can be grouped into distinct categories (some examples are shown in Figs. 2.1-2.4) [53,54,55,56,57]:

> ➢ desktop scanners - designed for the 3D scanning of small- to medium-sized objects;

> ➢ handheld, portable and wireless scanners are designed to be carried and operated by hand, and are ideal for obtaining 3D models of medium-sized objects ranging from human bodies to various technological items;

> ➢ professional and industrial scanners for precision in measurement and metrology - are designed so as to meet the high standards of precision and detailing needed for applications such as product design; the industrial scanners are used to capture images of small objects, or the intricate details of larger industrial objects, with high resolution;

> ➢ body-scanners and 3D scanning booths - are designed to scan entire human bodies or body parts (face, hands, feet, limbs).

For the purpose of taking body measurements, the indirect method can keep the person in a fixed position for a short time while performing the scanning. It allows the taking of plan or orthogonal images, from which the various body measurements can be calculated. The resultant data can then be integrated into a CAD system, and the information used for [44,48]:

- garment production,

- designing ergonomic spaces for workers or cars,

- designing sports products or divers devices for disabled people, and

- avatar-designing for computer games or virtual garment-fitting.

Some of the disadvantages of this method include increased equipment-cost and errors in curvilinear measurements which arise from shading effects during scanning and inconsistencies in body-positioning. As in the case of the traditional method, the scanned person must wear underwear or tight-fitting clothes, leading to intimacy issues. Challenges can arise in determining subtle features of the body [58]. The scanning method can be used to elaborate anthropometric standards for use in research with regard to the scanning of particular body-types or postures, or the tailoring of clothing.

*Figure 2.5. Various body-scans using Artec
handheld scanners, Source Artec 3D [59]*

3D scanning technologies are used to obtain data on various parts of the human body, or on the entire human body. Handheld scanners [59,60,61] are easy to use for scanning small parts of the body (Fig. 2.5). The entire body can be scanned using the whole-body scanners which are also available on the market. Due to developments in the technology, more precise and low-cost 3D scanners have recently appeared on the market [64]. A series of data obtained using body-scanners are shown in Figs. 2.6, 2.7.

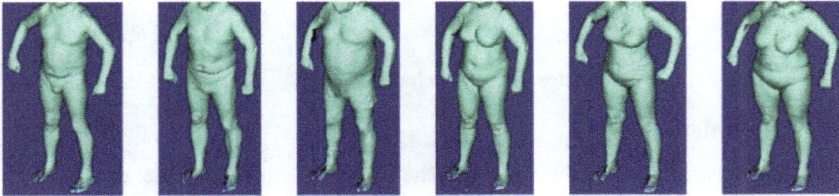

*Figure 2.6. Various whole-body scans obtained with <u>Vitus Smart XXL 3D</u> laser scanner from
Human Solutions [62]*

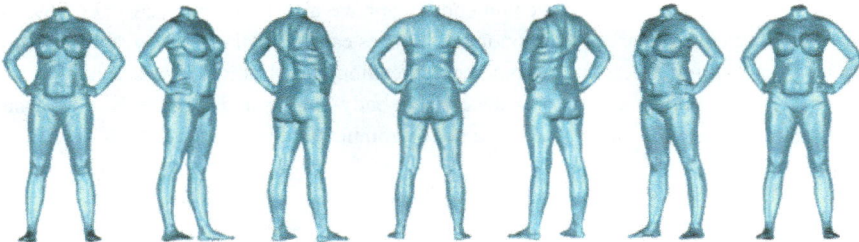

*Figure 2.7. Whole body scan of a female subject from different angles with the BodyLux® -System
from ViALUX, Source ViALUX [63]*

It can be seen that in order to obtain precise scanned data, it is necessary to know the scanning procedure or protocol and to specify a particular stance: upright position, with legs spread and hands slightly apart from the body. Only a few areas can then be missed by the scanner camera, and can be filled-in using the various software programs available for point cloud and mesh editing. The scanning data are finally processed in order to derive digital measurements of the surface of the body. The special software used for extracting these measurements (e.g., ScanWorkX, *Human Solutions*) permits the tracing of various plans and sections in order to deduce lengths, depths, widths, heights, angles-of-inclination and circumferences [48].

The main sequence used for obtaining 3D measurements is: Scanning › Data registering › Image processing › Feature extraction › Listing of the values of the measured dimensions (Figure 2.8).

Figure 2.8. 3D measurements extraction process

2.2   Measurement techniques for wheelchair-users

2.2.1   Traditional method

Anthropometric measurements of wheelchair-users are used for the design of their occupational environments and products. Many of the existing studies in the literature, regarding their anthropometry, were conducted with the aim of designing living, public or working spaces which eliminated harmful or uncomfortable situations. Accessibility demands and particular design recommendations have also been analyzed [67,68,69,70]. The measurements required for ergonomic purposes can be obtained in conventional ways by using measuring tapes, anthropometers, spreading and sliding calipers, cylinders etc. The necessary ergonomic data are deduced from body-lengths and perimeters. Wheelchair dimensions can also be taken in the case of environmental design (Fig. 2.9) [67,71,72].

Materials Research Forum LLC
https://doi.org/10.21741/9781644901557

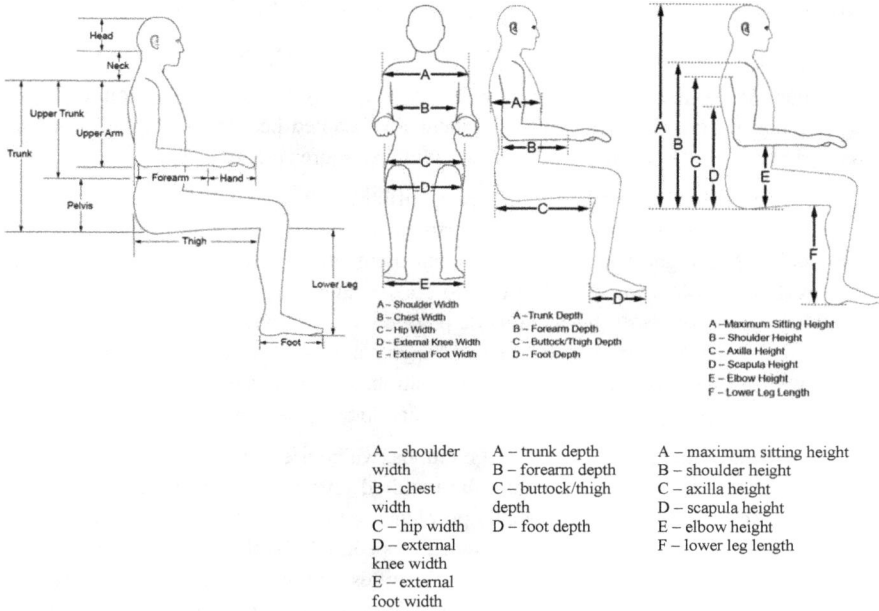

A – shoulder width
B – chest width
C – hip width
D – external knee width
E – external foot width

A – trunk depth
B – forearm depth
C – buttock/thigh depth
D – foot depth

A – maximum sitting height
B – shoulder height
C – axilla height
D – scapula height
E – elbow height
F – lower leg length

Figure 2.9. Examples of body segmentation (left) and anthropometric measurements of wheelchair-users for workplaces [72]

Different measurement procedures, using standard tools to obtain anthropometric data on wheelchair-users, are described in specialist publications. Some of these measurement tools are:

- the handheld Harpenden Anthropometer [73]; the ISAT Table was used to illustrate the person in a particular position for the making of direct measurements of the body;

- the meter stick and inclinometer to measure the linear dimensions needed to establish the spatial requirements of a person with disabilities [74];

- the Martin-Saller anthropometric instrument which uses a special chair that has an adjustable height and depth [67];

- the conventional tape-measure and ruler [69];

- calipers and anthropometers [70];

- the standard anthropometric chair;

23

- calipers in large and small sizes, tape- and metal-measures (1mm accuracy);

- stadio-meters (1mm accuracy) and scaled boards for anthropometry [75].

The importance of the product, the necessary accessibility to physical environments and the re-sizing of personal items and equipment for disabled people, has attracted further attention to the need to standardize measurement procedures [72,76,77].

Ergonomic studies of wheelchair-users are not sufficient for garment construction, because the ergonomic measurements are more pertinent to the environment within which the person is living. For garment construction, measurements have to be made directly on the body, as described in Table 2.1. The curved body features are measured by using a tape and following the shape of the body while paying particular attention to areas where the basic pattern has to be modified. In order to design a made-to-measure product which is in accord with the body-shape in the sitting position, it is also necessary to measure the dimensions of various body areas while in a dynamic body position [78,79,80].

The main disadvantages of conventional methods, which are also time-consuming, arises from the subjectivity of the results and the required immobilization of the subject for a relatively long time; this being especially problematic for wheelchair-users. In the case of traditional measurements, some of the standard specified positions are not adequate for disabled people, due to their various body deformities. Particular attention should be paid to the atypical distribution and use of muscles in both the lower and upper parts of the body, and to atypical bone mass or body positioning (Fig. 2.10) due to the disability [72,74].

Upon analyzing the measurement procedures with respect to the positions adopted by wheelchair users, it can be seen that it is not possible to specify a unique one for every person studied. While a person having a less severe disability can adopt a normal sedentary position and maintain a sequence of working motions, the situation is different for those suffering from a more severe disability. The measurements in these cases have to be taken in conformity with their normal positioning, with no aiding devices to support them [74,68]. In the case of the measurements needed to construct garments for disabled persons having a higher level of spasticity, it is necessary to lay that person on a bed in order to measure the hip circumference and the height of the body [80].

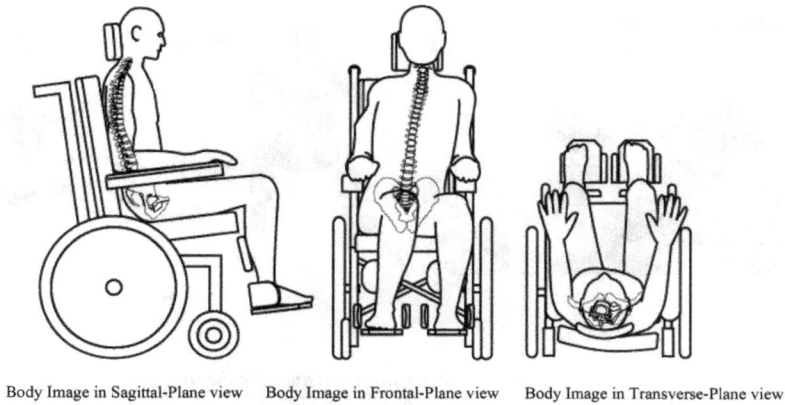

Body Image in Sagittal-Plane view   Body Image in Frontal-Plane view   Body Image in Transverse-Plane view

Figure 2.10: Postural deviations of a wheelchair-user in the sitting position [72]

Wheelchair-users have special needs which require optimization of their surroundings, combined with the functional aspects and usability of the garments and products which they use. This kind of optimization necessitates conducting an appropriate anthropometric survey that takes account of variations in body size, limitations on the demands of the wheelchair-user, variations in measurement procedures and atypical body positions in the sitting position which are due to variations in the afflicted area of the body [81].

2.2.2 Three-dimensional scanning procedure

Three-dimensional body-scanners are widely used for obtaining anthropometric parameters; usually in the standing position. The obtention of 3D measurements in the sitting position is still only in the research stage, it being difficult to choose the optimum position, equipment and scanner which can cover the entire body without missing some parts.

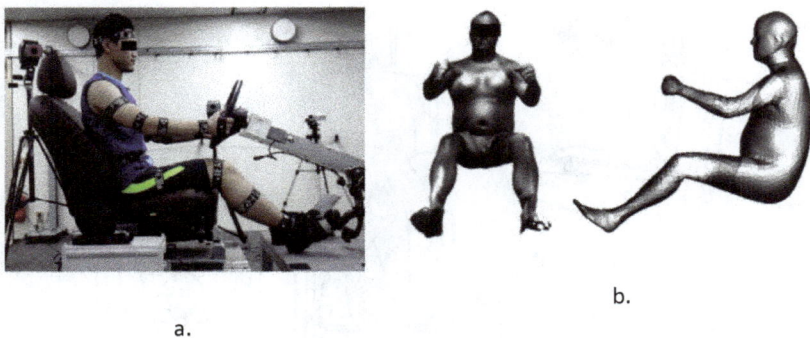

Figure 2.11. Scanning images of the sitting posture for the automotive industry (source: a.- [82],b.- [83])

Many ergonomic aspects can be analyzed with the help of scans of the sitting position. This technique is used in the automotive industry, where studies are made with the aim of improving the comfort of drivers and passengers occupying car-seats (Fig. 2.11) [82, 83] or replacing the conventional methods of ergonomic research in garment construction. It is difficult to obtain a complete scan in the sitting position because there is always a large section of data missing from the crotch and thigh area, the underarm area, the buttocks and the back of the legs [86]. The use of handheld or whole-body scanners (e.g., as in Table 2.2) is not sufficient to obtain accurate data on the sedentary position. It is usually necessary to repeat the scanning procedure so as to capture various areas of the body which can then be merged by using various modeling-programs (e.g., RapidForm and Geomagic from *3DSystems* [84], Polyworks IM Merge from *InnovMetricSoftware Inc.*[85]). After merging the various scans, it is necessary to carry out some mesh-repair; filling-in the holes and smoothing the surface so as to produce the final 3D form [83,86].

There are some disadvantages in using both handheld and whole-body scanners. The main disadvantages of using the handheld scanner are that:

- to obtain a 3D full-body scan it is necessary to have plenty of free space around the person so as to be able to perform a complete rotation,

- in the process of moving the scanner around the person, wires and equipment can become tangled or broken,

- because of the shaking resulting from hand support, the software can lose track of the image; thus necessitating re-positioning for a new scan, or even re-calibration,

- the scanning procedure takes too long in the case of full-body scanning, and

- the handheld scanner is more appropriate for scanning small areas.

The main disadvantage of the full-body scanner is that the overall size of the entire system is quite large, making the scanner more cumbersome to transport. The high price is also an impediment to the purchase of such scanners. The body-scanner can however be faster and more accurate: a full 3D body model can be obtained within a matter of seconds [87,88].

The scanning procedure for wheelchair-users has been described in some studies, but most work involves healthy people. The main aspects of obtaining a useful 3D body model in the sitting position are the necessary equipment and the scanning procedure which is to be followed. There is particular interest in acquiring an accurate 3D body model in the sitting position for the purposes of garment prototyping because, during the past few years, industrial interest in customized products has been tending to grow in tune with an increasing awareness of the demands and needs of disabled peoples.

Several methods for the scanning of the sitting position have been explored with regard to the equipment required, the scanning procedure or the body-posture required for the optimum gathering of the scanning data. According to some studies [89,90,91], the scanned person has to sit with the knees bent at nearly 90°, the hands slightly upraised from the body and with space left between the thighs. The clothing of the subject is also important and it is necessary, for optimum scanning, to wear light colors and close-fitting garments so that the light from the scanner can capture fully the contours of the body. Whole-body scanners such as the Vitus Smart XXL from *Human Solutions* [66], the Atos II 400 from *GOM* [92] and the Eva 3D from *Artec* hand-scanner have been used. Scanning of the subjects was performed from different angles and heights in order to get a complete image [93,94,95]. Having established the optimum technique, a trial scan of paraplegic subjects was performed [91,94]. The methods used are not a good solution for obtaining information from the sitting area; the buttocks and the backs of the legs are missing from the scan data. Clean-up and repair of the 3D body mesh are necessary following registration. Because of the large area missing from the body, it is difficult to follow the natural contours of the body, thus hindering acquisition of the correct measurements. Further studies aimed at finding a better scanning method are clearly needed.

2.2.3   Kinematic body-model

The animation of a kinematic body-model in a 3D program could be an interesting approach to observing changes in the human body when in the sitting position. The field of application of virtual human models is wide. The adaption and individualization of human models can be applied in areas such as [96,97]:

➢ ergonomics,

➢ the entertainment industry (e.g., film and video production, games, sports),

➢ medical rehabilitation,

➢ product design (e.g., automotive industry, virtual clothing construction and fitting),

➢ training via interactive simulations, and

➢ information provision (e.g., museums, airports, websites).

A kinematic body-model is the result of merging scan data with a template model, within a body-animation program, with the aid of defined reference-marks. Within such animation programs, body motions can be controlled and the change in body shape during such motions can be analyzed [98,99]. Special attention in virtual modeling is paid to deformations of the skin and muscles during movement, and to the variety of human body morphologies [100]. Graphic software such as Maya or 3dMax offer solutions for the drawing and animation of the human body in static and dynamic positions [101]. The process of producing animations of a kinematic body can involve two main approaches: *anatomical animation* and *data-driven animation*.

Anatomical animation

Anatomical animation, also called *skeleton-driven animation,* is the standard method for putting human scan data into motion by connecting it to a template model (the latter may be just a skeletal system, or it may be built up from a skeletal or muscle system which is attached to a skin surface) [102,103]. A personalized kinematic body-model which can be animated is obtained by adapting the template to the scan data (Figs. 2.12, 2.13). *Skeleton-driven animation* is integrated into most animation software and is commonly used to analyze various ergonomic aspects of the body during diverse motions.

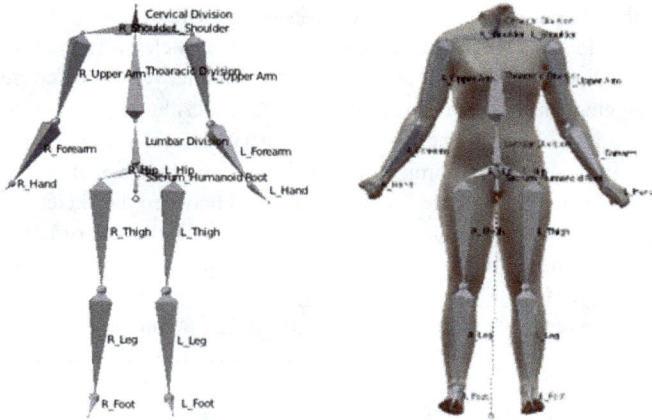

Figure 2.12. Skeleton template (left) and skeleton template attached to scan data (right) [93]

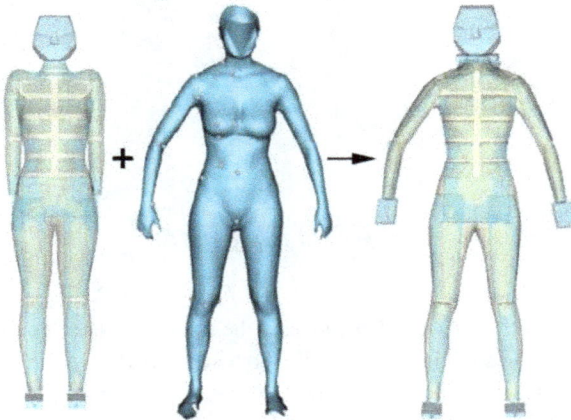

Figure 2.13. Template model (left) and scan-data fitting [99]

The 3D body motion in skeletal animation usually consists of repeated movements of the joints and limbs. These movements are described mathematically by the consecutive coordinates and orientations of the body parts [104]. "The skeleton is a set of rigid links connected by mechanical joints, while the muscle-tendon network is composed of wires representing the elements to drive and/or constrain the bones, including muscles, tendons, ligaments and cartilages" [105]. The skeleton is not usually presented realistically as it only plays the role of providing information concerning body deformations for specific movements and positions. A muscle layer can be added between the skeleton system and the skin surface so as to create a more complex template model that can offer better surface deformation during the animation process. The latter can be done by using various programs (e.g., MayaMuscleCreator, or 3dsMax from *Autodesk* [106] - Figure 2.14). The muscle layer takes over the deformation of the skin surface at joint locations [97,107].

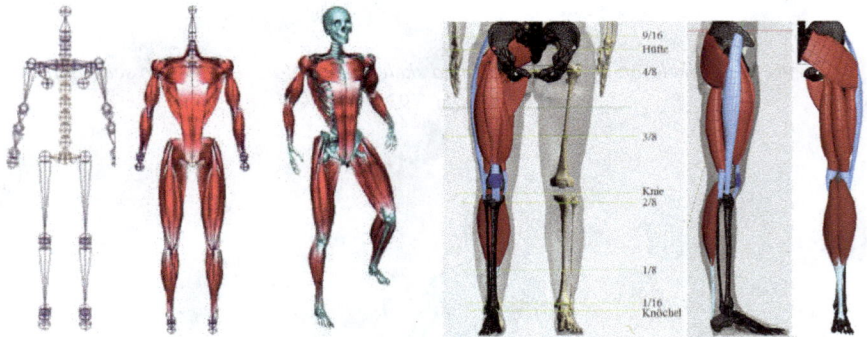

Figure 2.14. Muscle structure in Maya MuscleCreator[a] [107] and the muscle structure of the right leg in 3dsMax[b] [97]

Another critical aspect of the template model is the skinning procedure, which connects a moving skeleton to the surface model, leading to a nearly perfect deformation of the skin. The commonly used skinning techniques involve a linear blend or dual quaternion skinning. The skinning technique for skeleton-driven animation is to deform the articulated character by using standard skeletal sub-space deformation [108,109,110].

The process for creating a complex template model which comprises a musculoskeletal system can have the following principal characteristics [102]:

> ➤ a suitable surface mesh derived from scan data,

> ➤ an anatomically well-defined skeletal system,

> ➤ a motion test-sequence,

> ➤ a muscle system added to the skeleton, and

> ➤ adjustment of skinning parameters to the musculoskeletal system.

In order to animate the kinematic body-model which is obtained by fitting scan data to the template model a final rigging process has to be performed. This rigging process involves [108]:

> ➤ skeleton blending - adjusting the position and orientation of the skeleton joints so as to fit the scan data,

> ➤ skinning binding – applying to the surface mesh of the template model suitable skinning algorithms and refining attributes (e.g., weight) in order to obtain an acceptable deformation of the surface, based upon a realistic deformation of the scan data, and

> ➤ setting controllers - designing manipulators which control the kinematic body-model.

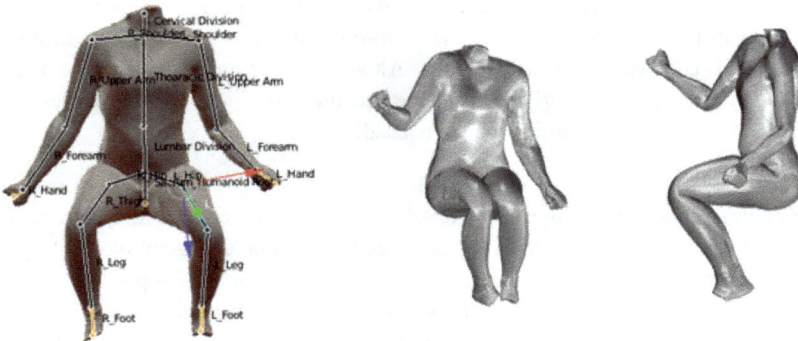

Figure 2.15. 3D body model adapted to various positions [93]

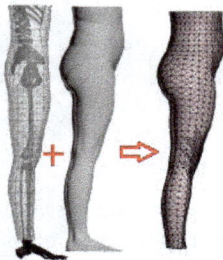

Figure 2.16. Template and female scan data for the lower part of the body [97]

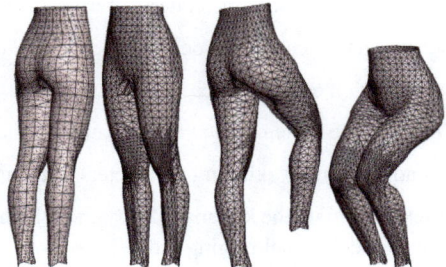

Figure 2.17. Various motions of the obtained kinematic body-model [97]

The rigging process is achieved by using a set of scripts and commands in the programming language. Scripts can be used to find errors in the template surface geometry, fix errors in the rigging process or control the animation [111]. Following accommodation between the scan data and the template model (skeletal or musculoskeletal template), the kinematic body model can be animated (Figs. 2.15 – 2.17). Various body postures can be set up, and the deformation of the mesh surface can be analyzed. The process of obtaining various postures by animating only a kinematic body-model with the help of a merged skeleton structure leads to some irregularities in the mesh-transformation and thence to differences between real and virtual measurements. Although it is an excellent method for understanding the problems arising from the sedentary position, further studies of the adaptive 3D model mesh-modification are needed.

Data-driven animation

In *data-driven animation*, also called *example-based animation*, the relationship between the template and the scan data is made directly, without a musculoskeletal system [102,108]. Both the template and the scan data in this case require corresponding reference points and the same initial position. The scans can be performed either by scanning the body in diverse positions or by capturing a motion sequence of the body in order to collect data. The obtained scan positions are further used on the template in order to adopt the same motion-sequence while calculating the changes that the surface mesh has to undergo between the postures adopted. Various different approaches are reported for modeling a virtual body, based upon the scan data obtained for diverse body postures and shapes. An example-based method is used in [112] for calculating skeleton-driven body deformations.

A kinematic skeleton was constructed by registering the articulated body deformation, for the upper part of the body, using a set of scan data for various postures.

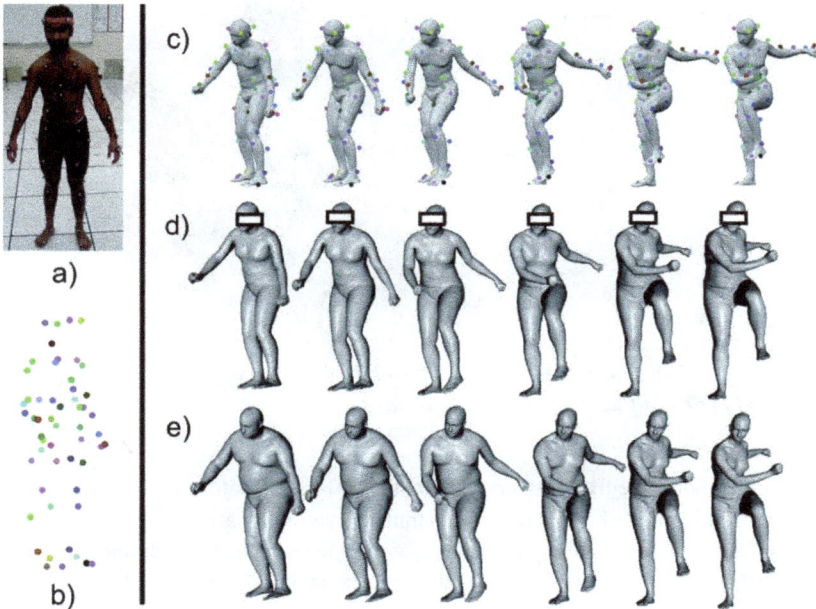

Figure 2.18. Motion capture for SCAPE model - a) scanned subject with motion markers; b) frame of the motion markers; c) animation of the scan data based upon the motion capture sequence; d) motion transfer to other scan data; e) motion capture with changed body-shape parameters [113]

The SCAPE method [113] led to a high-quality animated surface model for a moving person, with realistic muscle deformation, using a single static scan and a marker motion capture sequence for the subject (Fig. 2.18). The method described in [114] was an attempt to design a model, with improved robustness plus minimal noise or missing scan data, which captures how body-shape varies across a range of people and postures. The idea was to derive, from different alignments obtained for various postures, a new kinematic body-model that could handle a new position that did not exist in the scanning sequence (Fig. 2.19).

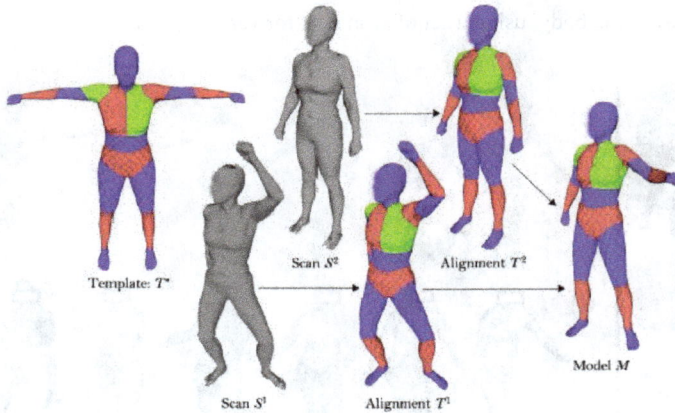

Figure 2.19. Alignment between a template and various scan data so as to obtain an articulated Model M that can adopt new postures [114]

A 3D human body, TenBo, was designed which offered variations in posture and body-shape. This method tried to improve mesh transformation in various postures by applying a tensor-decomposition technique. The newly designed kinematic body-model proved to be better than the SCAPE model for obtaining realistic body-shapes [115].

The scanned data in all of the studies presented were obtained for static positions. Another approach [116] was presented for obtaining a virtual body-model from scan data for various positions. The Dyna model can follow soft-tissue deformations on the basis of previous deformations which are obtained from a motion sequence, the velocity and acceleration of the body and the angular velocities and accelerations of the limbs. This method also determines how the deformations vary as a function of the body mass index (BMI) of the person; predicting differing deformations for people having different shapes (Fig. 2.20). Although the *data-driven* methods can offer animatable 3D models having realistic body-shapes, there still exist problems regarding the mesh transformation for various postures.

The transition from one position to another (standing-sitting) cannot be accomplished without involving intermediate states. The shading problem for various areas of the body during motion-capture still intrudes. Another problem with this method could be the significant volume of work involved in gathering scan data for just one pose and only one body-shape in motion; not to mention the inevitably expensive equipment and software.

Figure 2.20. Dyna model – template body-mesh(blue), aligned to scan-data (red), and representations of various animated body-shapes in motion (grey) [116, 117]

2.3 Computer-aided design systems

Garment CAD technology is the use of computers to design clothing products [118]. The CAD programs for garment construction used 2D methods in their early development, including pattern-design, pattern-grading and marker-making. The CAD technology developed over time, and 3D garment-construction attracted attention by being able to achieve a better understanding of the fit of a garment, well before physical construction of the product.

The technology was then intensely studied and developed due to its importance to the garment industry, with the primary research being focused upon garment-modeling and simulation, garment design, garment-grading and the availability of 3D CAD systems to the clothing industry [119,120,121,122,123,124]. These reviews summarize the information pertaining to virtual prototyping, a process which can be approached using both 2D-to-3D and 3D-to-2D techniques. Virtual prototyping significantly shortens the times, and lowers the costs, involved in product design. [125].

2.3.1   2D-3D virtual prototyping of garments

The 2D design techniques are based upon several sizing rules, using standard body measurements and conventional garment-design methods. There are several CAD programs on the market which are used by the clothing industry for pattern making, such as AccuMark– *Gerber* [126], Gemini Pattern Design – *Gemini CAD Systems* [127] and Modaris – *Lectra* [128]. The 2D technology has the following main characteristics [129]:

- the advantage of being an established process, thus making it more acceptable for garment design and construction,

- can be used separately, without 3D visualization,

- requires skill in, and great experience of, traditional pattern-making, and

- lacks the ability to visualize the product before its physical construction.

The garment industry is based mainly upon 2D pattern-making for product manufacturing but, more recently, 3D virtual fit simulation has started to attract attention. Virtual fitting to 3D body-models is an essential step in reducing prototyping during garment development [119,129]. The basic 2D patterns can be designed, and later assembled, via a virtual sewing procedure so as to achieve realistic draping of a simulated 3D mannequin (Fig. 2.21) [130,131,132]. Modeling of the behavior of the fabric is essential in order to ensure realistic simulation. The mechanical properties of fabrics can already be chosen from an existing library of fabric simulation programs. A CAD system for the clothing industry should have [133]:

- a fabric library which can easily provide the properties required for drape simulation,

- a virtual body-model which can be adapted to various body-sizes, and simulations for diverse patterns which can exploit fabric-library properties.

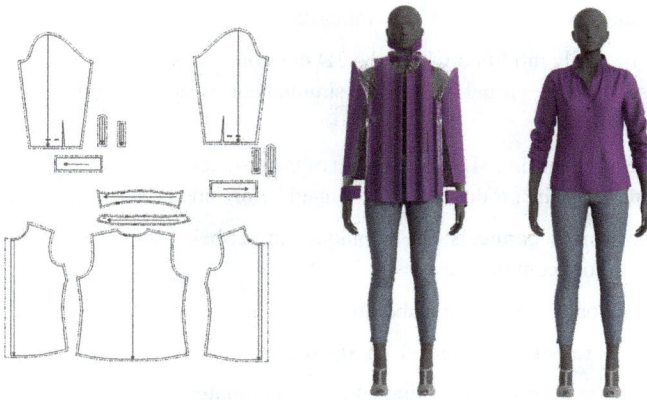

*Figure 2.21. Virtual fit simulation of 2D patterns to a virtual body-model, using Assyst software,
Source Assyst [47]*

2.3.2   3D-2D virtual prototyping of garments

The 3D virtual garment construction technique used in clothing development requires innovative CAD solutions. Virtual prototyping aims to integrate garment characteristics, pattern cuts and fabric properties so as to be able to check the fit on a virtual body-model [122]. It is possible, within this 3D environment, to modify the shapes of garments and to introduce specific fabric properties.

The 3D-to-2D technique is based upon developing 3D human-body measurements and modeling, 3D garment design using virtual body-models, 3D garment simulation and 2D pattern-generation based upon the resultant 3D garment model [119,120]. A flow-chart for the 3D garment-design process has been constructed [130]. The process starts with the development of clothing by using the 3D mannequin; a flattening procedure is then applied in order to obtain 2D pattern-pieces. The process may continue with a 2D-to-3D procedure in order to detect any changes to the pattern-pieces that occurred during the flattening process.

The efficiency of the 3D technique can be summarized as follows:

- the patterns can be easily modified within the 3D environment so as to adapt them to the shape and dimensions of the virtual body while simultaneously analyzing the changes that occur to the 2D patterns,

- has the advantage of permitting 3D visualization of the product as fitted to a virtual human body; the shape and fit of the product can be changed before creating a physical prototype

- is used mainly for simple garments; the technique not yet being well-enough developed to be able to handle more complex designs,

- makes designer and pattern-maker collaboration more necessary,

- allows simultaneous real-time changes from 3D to 2D,

- demands a knowledge of software-writing and other computer skills,

- requires a virtual library of the fabrics and textures to be used for garment construction.

The 3D garment-construction technique can be used for both tight-fitting [134,135] and loose-fitting garments [136,137]. In the case of tight-fitting garments, the pattern lines and curves are drawn directly onto the surface of the 3D body-model while, in the case of loose-fitting garments, it is necessary to use a 3D construction algorithm for a second skin which will treat the garment as having various clearances around the 3D body-model [138]. Various studies have been carried out with regard to tight-fitting and loose-fitting garments and the development of parametric 3D body-models that can be used as mannequins in 3D design programs [139,140,141,142].

Most 3D garment design studies are aimed at industrial clothing manufacturing, which assumes standard body-types for certain groups in the population. The use of 3D garment design is problematic for the manufacture of individual clothing items that takes account of the specific body characteristics of an individual. The use of 3D design has recently attracted increased attention, in the clothing industry, with regard to persons having special needs. Studies of virtual garment-prototyping have started to analyze various fitting problems which arise from the position of the body [144] or from disabilities such as scoliosis [145,146] and paraplegia [89,94,91].

2.3.3   Conventional garment construction for wheelchair users

When creating customized garments for wheelchair-users there are many dimensional changes, with respect to the basic model, that must be considered. Several studies have focused on the pattern modifications which have to be made to the lower and upper pieces of this type of clothing. These studies show that the lower piece of the clothing for a body usually requires modification in length and waistline. That is: the crotch is shortened at the front and lengthened at the back. Extra darts are added at the waistline so as to even up the difference between the hip area and the waistline when in the sitting position [78]. Other modifications which have to be made are to lengthen and widen the pants, re-position the knee-line and widen the sedentary crotch area [89]. In addition to pattern modification the placement of the opening system was reconsidered by adding special plackets at the sides, plus zippers for detaching the crotch area or elastic panels in the waistline to facilitate dressing and undressing [147,148]. Detachable parts or tight-fitting bottoms were also considered; using soft and water-absorbent knitted fabrics, ones made from a mixture of synthetic and natural fibers, softer materials at the back and flat smooth seams [42]. The upper part of the body usually needs to have reinforced sleeves, a longer back, a shortened front, large armholes, full sleeves without cuffs and a front having a central placket which is easy to use (Figure 2.22). Another example is the elbow area, which needs more darts and pleats. Special attention to the placement of the opening system is needed [42,89,147,148]. As well as normal clothing elements, the use of protective textile products also demands attention; such as protective pads or diapers for managing incontinence and imparting resistance to wind and rain [41,147,149].

Standard modifications to be applied to customized clothing for wheelchair-users have not yet been formulated because each study has considered an individual case. The modifications and parameters which need to be applied in pattern construction can vary due to the particularly unique features of the body of a paraplegic, for example. Disabilities govern the special requirements with regard to functional clothing, or to any other textile products which are aimed at improving the lives of the disabled. The design recommendations which have been proposed so far do not cover the complete and unique customization of a garment. They offer however an overall view of the basic pattern-modifications that can satisfy the particular requirements of wheelchair-users with respect to clothing design; including pattern construction, fabric characteristics and aesthetics.

2.3.4   3D virtual prototyping of garments for wheelchair-users

A series of studies were made using the 2D-to-3D technique for the virtual prototyping of garments for paraplegics. The research was carried out using OptiTex 3D [89, 91, 95]. The patterns [89] for a basic pair of pants and a blouse in the upright position were designed and virtually simulated for a sitting body-model. By using the virtual dimensions which were obtained from the sitting body, the patterns were then drawn and new ones virtually simulated (Fig. 2.22). In order to reconstruct the patterns for the pants, the length was increased, the knee-line was lowered, the front of the pants was also lowered, the rear crotch-length was increased and the widths of the knee-area and leg were increased. All of the changes to the patterns were made in 2D, while carefully noting the dimensions of the 3D body-model and tensions in the fabric which arose from the fit-simulation. This procedure requires good pattern-making skills and a knowledge of garment design. In the reconstruction of the blouse, intermediate virtual fittings were required in order to monitor the necessary changes. The front of the blouse was therefore modified so as to cover the tights while sitting. On the basis of this virtual prototyping of the pants and blouse, it was found that the basic pattern design did not correspond to a sitting position. The proposed modifications to the basic patterns represent the first step in understanding garment construction for the sitting position, but the procedure is still cumbersome and requires a different approach.

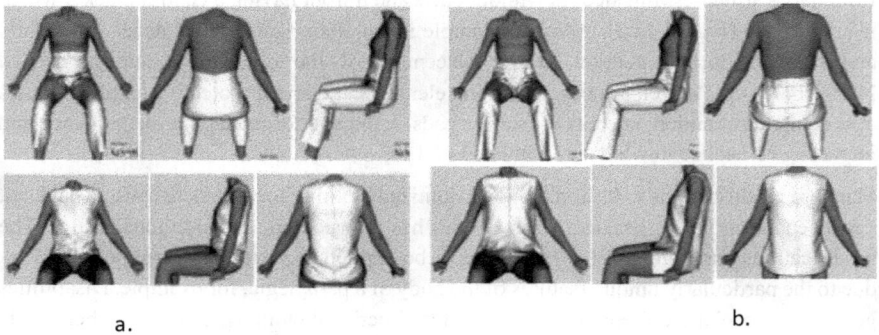

a.   b.

Figure 2.22 Fitting of pants and blouse before modification (a), and after modification (b) [89]

One study [91] has described the development of three garments for wheelchair-users: a jacket, a dress and a pair of pants (Figs. 2.23 - 2.25). The virtual prototype of a garment was based upon virtual measurements taken from the scanned body of the disabled person. Virtual fit-simulation of the model design was subsequently carried out for standing and sitting positions. A comparison of the virtual fittings made possible an analysis of the modifications of the garment which were required for the sitting position.

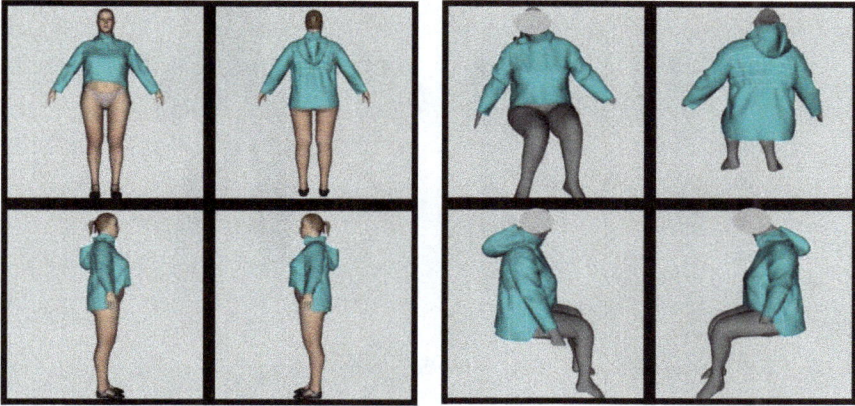

Figure 2.23. Virtual simulation of the customized jacket on the standing and sitting mannequin
[91]

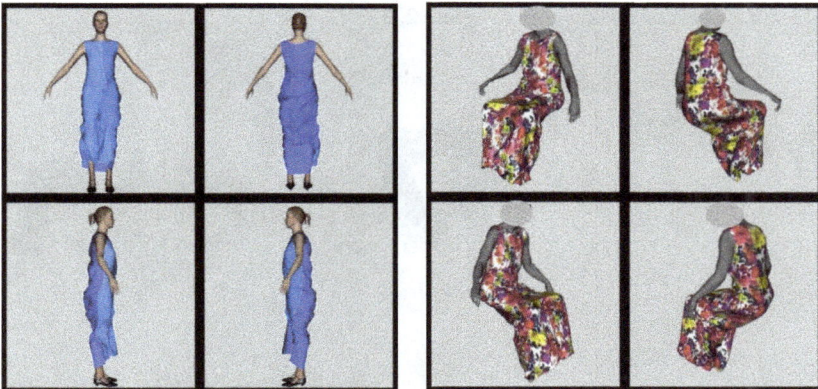

Figure 2.24. Virtual simulation of the customized dress on the standing and sitting mannequin
[91]

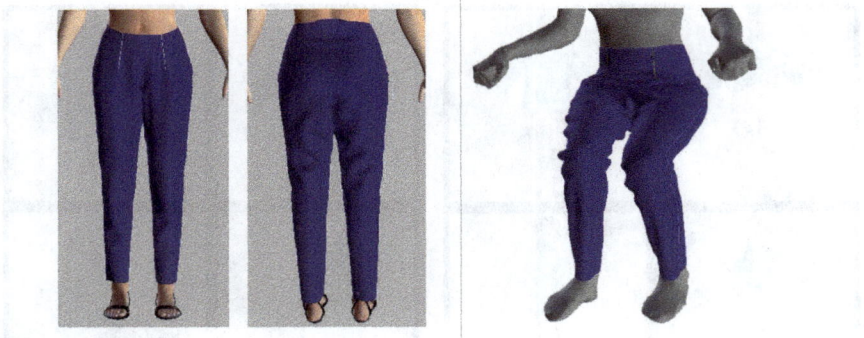

Figre 2.25. Virtual simulation of the customized pants on standing and sitting mannequins [91]

Attention in another paper [95] was focused upon developing functional pants which were adaptable to individuals who are prone to suffer from pressure-sores, incontinence or sweating in the hip and crotch area, between the thighs and under the buttocks. Particular antimicrobial and anti-oxidative fabrics were anticipated to be used for these problematic areas (shown in grey in Figure 2.26).

Figure 2.26. Pants with antimicrobial and anti-oxidative textile materials, adapted for wheelchair-users [95]

This literature-survey of the 3D scanning and virtual prototyping of garments for wheelchair-users gives a basic idea of the individualization of functional garments from the ergonomic point of view and better helps to understand the functional and aesthetic needs of paraplegic people. The problem with the scanning procedure is the fraction of data

Materials Research Forum LLC
https://doi.org/10.21741/9781644901557

which is missing from the scans; namely the areas, missing from the buttocks and pelvis, that need to be reconstructed. To follow closely the natural shape of the body is difficult and it is probable that some of the virtual measurements are imprecise, as acknowledged by these studies. The proposed virtual prototyping method is time-consuming because, for the purposes of pattern modification in particular, it is necessary to repeat the measurements several times in order to treat the scanned body in the sitting position. The basic patterns need to be modified more than once, to a degree which depends upon the particular body dimensions of a given person. The new models also require several intermediate virtual fitting simulations to be performed until the final model is obtained. Each garment is individually designed for the particular needs of a specific wheelchair-user, no general pattern-making method having been established. There is thus still a need to improve the scanning procedure for the sitting position in order to obtain more data from the contact area and thereby obtain better 3D body models for the making of accurate virtual measurements and optimizing virtual prototyping. The virtual prototyping of garments would be less time-consuming if the 2D-3D technique could be transformed into a 3D-2D technique as it would then be easier to see, from the very outset, the dimensional modifications which had to be made to the basic patterns.

The significant lack of dimensional data, which are required for garment construction on behalf of wheelchair-users, was therefore the starting point of the work to be presented in the following chapters. The proposed scanning procedure and virtual-garment prototyping are the best hope for obtaining improved results.

Chapter 3

3. Proposed Methodology for Body-Scanning

The development of any type of clothing product is based upon the shape and dimensions of the body. Clothing for persons with special needs has to allow more freedom and independence in dressing and undressing, to provide adequate fitting of an atypical body-shape and to have an excellent fashionable appearance. Garments for paraplegic persons should satisfy all of those particular needs as well as offering ergonomic comfort in the sitting position. With regard to the needs and demands of paraplegic people concerning clothing, 3D scanning procedures and computer simulation techniques were analyzed in order to explore the possibility of using a kinematic body-model in a virtual garment simulation prototype dedicated to wheelchair-users (Fig. 3.1).

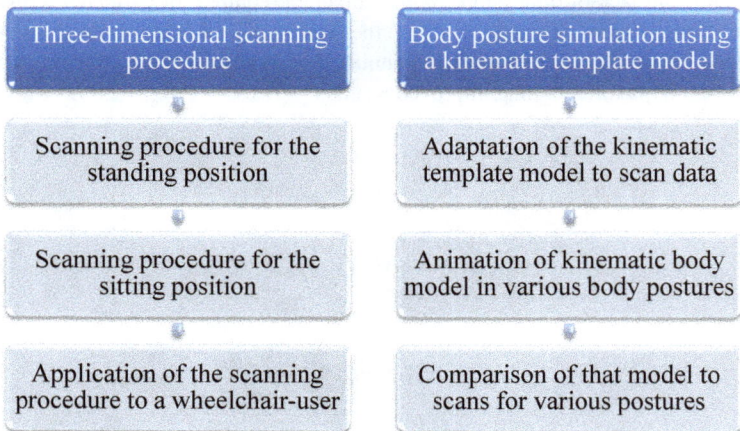

Three-dimensional scanning procedure	Body posture simulation using a kinematic template model
Scanning procedure for the standing position	Adaptation of the kinematic template model to scan data
Scanning procedure for the sitting position	Animation of kinematic body model in various body postures
Application of the scanning procedure to a wheelchair-user	Comparison of that model to scans for various postures

Figure 3.1. The steps followed in the study

The procedure was carried out by using the handheld *MHT* scanner from *Artec* [150] and the *zSnapper* body-scanner from *Vialux* [151]. Both scanners were used to assess their capabilities with respect to obtaining good scan data for virtual body-models. Editing of the mesh surfaces of the resultant scan data was carried out using the _GeomagicStudio_ program from _3D Systems_ [152]. The scanned image of the body was imported and animated using the _3dsMax_ program from _Autodesk_ [153] in order to obtain various positions of the lower part of the body. This entire procedure was necessary for the optimization of pants-design in the sitting position, with the aim of offering suitable ergonomic comfort to wheelchair-users. Inadequate fitting in length or waistline, and folds in the fabric around the bent legs or in the buttocks area, usually lead to health problems arising from inappropriate thermal insulation, impaired blood circulation, skin irritation or pressure-sores. The importance of suitable garment-fitting therefore has not only an aesthetic aspect but is also vital to the health of paraplegic people.

The 3D scanning procedure was carried out on three participants: two healthy females and one paraplegic male. The use of healthy people in the first stage was necessary in order to establish a scanning protocol for the sitting position, and to select devices that could best aid the scanning process. Because of the long periods of time required to accomplish this, it was unreasonable to expect a paraplegic person to remain immobilized in the necessary manner. It was also necessary firstly to develop, as a matter of methodology and time-management, a scanning procedure that would not deter a wheelchair-user from accepting it. The trials which were performed using healthy participants helped to understand better the usability and efficiency of both the handheld and whole-body scanners, to choose the methodology to be applied so as to obtain the best data under better time-management and to arrange the optimum set-up for scanning of the sitting position.

3.1   Three-dimensional scanning procedure

The three-dimensional body-scanning principle is a non-contact method which is based upon the optical triangulation of surface points by using laser or conventional light beams. In optical triangulation (Fig. 3.2) a beam of light is projected onto the object to be measured, and the reflected light is captured by a sensor [154,155,156,157].

The archetypal scanning system therefore comprises light/laser sources, image-capturing devices, software for analyzing the collected data and a computer which has a monitoring system for visualizing and processing those data [158]. All scanners use the projection of light patterns onto the area which it is required to scan. The equipment also includes compatible software for communication between the scanner and the computer.

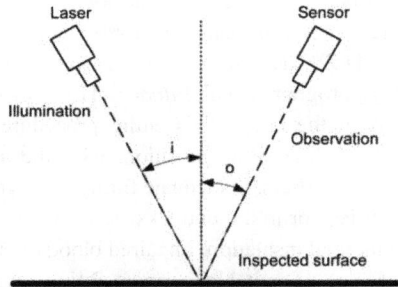

Figure 3.2. Principle of triangulation, where i is the incidence angle, and o is the observation angle [159]

3.1.1 Scanners

The *Artec MHT* scanner (Fig. 3.3) [150] works on the principle of projecting a reference light-pattern onto the object which it is required to scan, and recording it using a camera. The 3D shape of the object is deduced by calculating the perceived deformations of the reference pattern. The *MHT* scanner can also capture color and texture, and work in video-mode so as to capture 3D color-data. Its high measurement rate gives the *MHT* scanner a better data-acquisition speed than that of laser scanners, as well as offering high-resolution and accurate scanning. The scanner is also equipped with software for data acquisition and processing: *ArtecStudio* is an Artec program that can be used for scanning using various scanners.

Figure 3.3. The MHT Scanner from Artec, Sourse Artec [150]

The steps required to acquire scan data are [160]:

- activation of the software and scanner, preparation of the object and choosing the scanning procedure: rotation direction, single-scan or multiple-scans for larger objects;

- cutting out surroundings irrelevant to the scanned object, aligning in the case of multiple scans and global registration so as to optimize frame-positioning for all scans;

- global registration to eliminate noise, to fuse and erase existing flaws, to simplify the mesh and mapping of the texture if required;

- scan-data measurement, and exportation in a suitable format for further processing.

The *VialuxzSnapper®* (Fig. 3.4) is a full-body scanner that collects data by rotating the subject through 360° and, as well as permitting 3D surface-scanning, it can also provide body measurements. The scanner uses a Digital Micromirror Device (DLP) and phase-encoded photogrammetry technology which offer good accuracy, speed and reliability [63]. The DLP is based upon a Micro-electromechanical System (MEMS) known as a Digital Micromirror Device (DMD) which consists of many small mirrors that project light under electromagnetic activation. The system offers very high contrast and a high light output [161].

Figure 3.4: The BodyLux® -System from ViALUX, Source ViALUX [63]

Photogrammetry is the technique of obtaining reliable information and measurements, concerning physical objects and the environment, from photographic images. The process is carried out by recording, measuring and interpreting overlapping images and patterns of electromagnetic energy [163,164]. The *VialuxzSnapper®* scanner can be used in many domains, from health-care to body-shape tracking for the sports, medical and clothing industries.

The scanning system consists of the scanner support, laptop and turntable. The person to be measured has to stand on the turntable while the software continuously measures and combines scan data, on the different views resulting from a complete 360° rotation, within less than one minute. It is possible to perform both full-body scanning, or cover just the lower part of the body (Fig. 3.5).

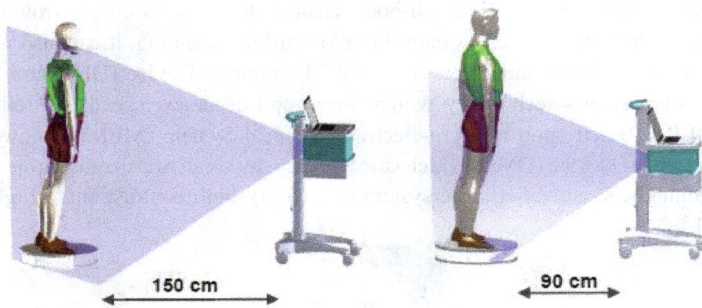

Figure 3.5. Full-body and lower-body scanning with BodyLux® -System from ViALUX [151]

The turntable stops automatically when the 360° rotation is complete, and the software then generates the 3D body-model. Measurements can now be taken or the data can be exported for the purposes of surface mesh modification or other processing. The accuracy and high speed of the scanning procedure make the ViALUX scanner a reliable method for acquiring scanning data in various domains.

3.1.2 Scanning procedure for the standing position

Some basic requirements have to be established, with regard to the scanning procedure in the standing position, in order to ensure optimum data acquisition: the scanner must be calibrated for reliable data acquisition, a minimal number of scans should be used in order to reduce the time involved, a maximum area of the body surface should be captured and the subject should be positioned so as to minimize the fraction of surface which is missed by the scan. When making anthropometric measurements in the conventional manner the standard posture is with the legs next to each other and the hands slightly away from the body. The optimum posture for scanning however requires that the legs should be more widely separated and that the hands should be even further away from the body. This positioning is necessary so that the scanner light can capture accurate data from problematic regions such as the armpits, and between the thighs [164]. The efficiency of the VialuxzSnapper® is particularly notable when scanning a healthy person in the standard position. The subject has to remain steady on the turntable, with legs separated to shoulder-width, and with the hands slightly away from the body. He has to maintain the same position while the turntable is rotated; this not being too difficult given the short time that is involved (Fig. 3.6). The subject has to wear light-colored tight-fitting clothes so that the light from the scanner can capture correctly the shape of the body. The room should be shaded, with no direct light from the Sun or from artificial sources.

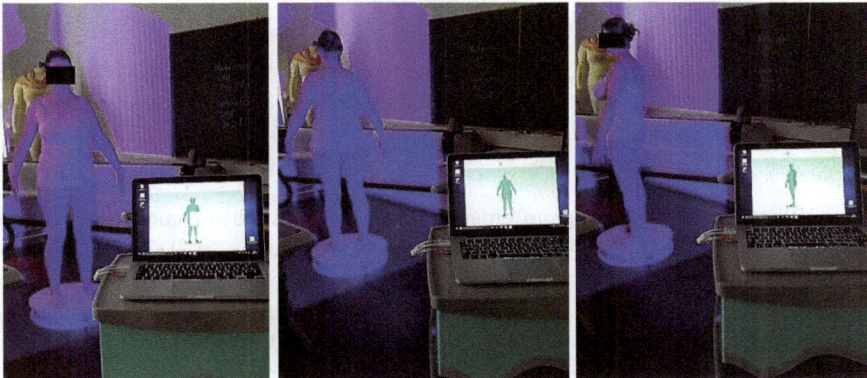

Figure 3.6. Body-scanning with the BodyLux®. zSnapper® - from ViALUX

The scan data which are obtained prove to be of good quality because the *ViALUX* scanner can cover the entire body using a single 360° rotation, and there are no overlaid or double images. Any parts missing from the inside-leg, crotch, armpit, hand, or shoulder regions are due to the inability of the scanning camera to see those shaded areas of the body (Fig. 3.7).

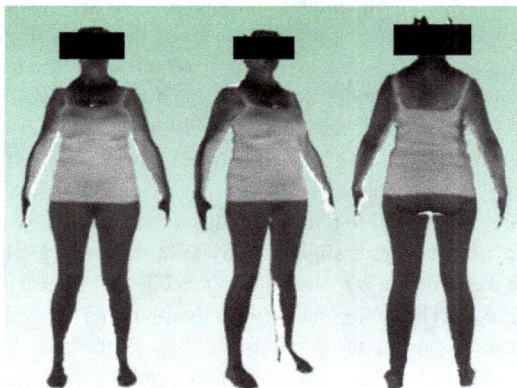

Figure 3.7. Scan images obtained using the BodyLux®. zSnapper® - from ViALUX

The data can be saved in *.obj* format and imported into <u>*GeomagicStudio,*</u> where they can be repaired by applying some basic mesh-editing (delete, trim, fill-holes) or advanced polygon-editing (sandpaper, patch, sculpt).

Surface reconstruction

The *3DSystems -* <u>*GeomagicStudio*</u> program was used for surface reconstruction of the Vialux scanner data. This software offers solutions for the editing of 3D scan images and their transformation into digital models which are suitable for use in reverse engineering, product-design or prototyping. The combination of mesh editing, cloud points processing and advanced surface functions helps to create accurate solid models for 3D applications. A polygon (also known as a triangle) is a plane which is formed by three line-segments. A polygonal object consists of polygons which are arranged into a complex mesh. In the polygonal phase (Fig. 3.8), the mesh can be automatically detected and repaired by using the *mesh doctor* tool. Self-intersecting and highly creased edges (an edge with a sharp angle between the normal vectors of the two neighboring polygonal faces), spikes and small clusters of polygons (a group of small isolated polygon meshes) can also be fixed. Other mesh-editing tools such as *smooth polygon mesh, define sharp edges, defeature holes* and *fill holes* are also provided in order to help users to create and modify polygon meshes [165].

Figure 3.8: GeomagicStudio- Polygonal editing bar

The first step in reconstructing the mesh surface of scan data for the standing position consists of filling the holes in the body areas where the scanner could not capture data (Fig. 3.9). The *Fill holes* tool fills gaps in a polygon object that are caused by sparse underlying point-data.

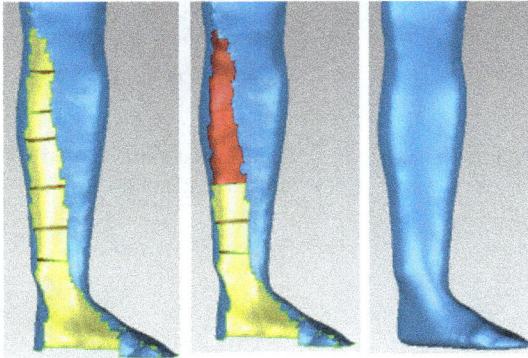

Figure 3.9: Hole-filling procedure

In places where the hole is too big the *Create bridge* tool can be used to obtain an anatomically excellent body shape. Following the use of the *Filling holes* procedure, the mesh surface may have to be improved. By using the *Remove spikes* tool, single-point spikes can be detected and flattened while the *Reduce noise* tool helps to align the points of the obtained mesh so as to give smoother curves. The *Relax* tool is finally used to smooth the polygonal mesh by minimizing the angles between individual polygons (Fig. 3.10).

Geomagic thus provides excellent mesh-surface processing tools which are suitable for filling-in, in an easy manner, areas which are missing from the scan data. Surface improvement of polygon mesh can also be achieved within a couple of easy steps: *Remove spikes › Reduce noise › Relax.* Performing these mesh-processing techniques leads to a final data-set which represents a realistic body-shape for the standing position (Fig. 3.11).

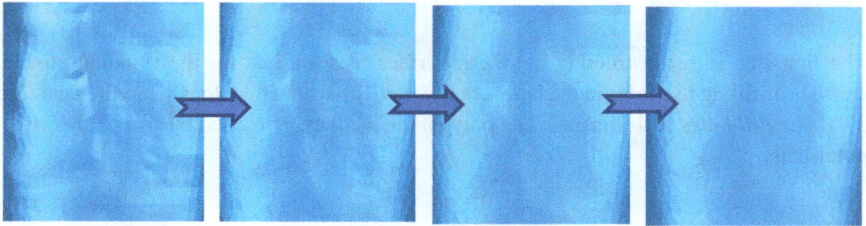

| a. Raw polygon mesh | b. Polygon mesh after *Remove spikes* | c. Polygon mesh after *Reduce noise* | d. Polygon mesh after *Relax* |

Figure 3.10. Processing polygon mesh using GeomagicStudio

When a good-quality mesh has been generated, the data can be used in the polygonal phase or a different *Shape phase* can be used to create NURB (Non-Uniform Rational B-Splines) surfaces.

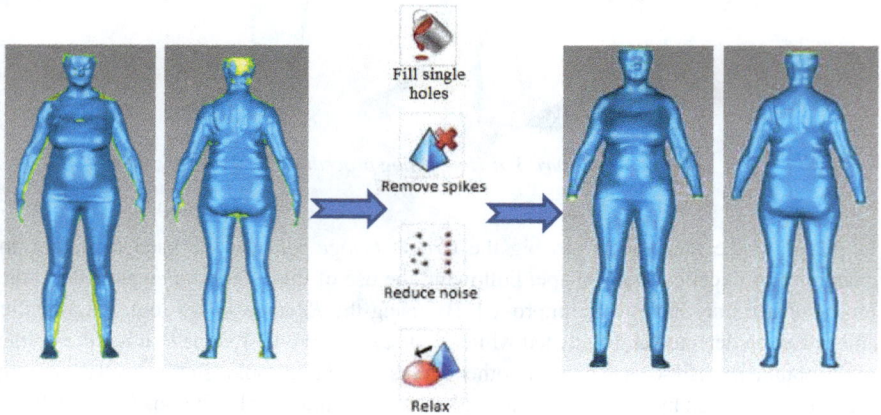

Figure 3.11. Polygon mesh-editing using GeomagicStudio

3.1.3   Scanning procedure for the sitting position

The main objective of the proposed scanning procedure was to explore the possibility of scanning the body in the sitting position. A larger area of the surface of the body is shadowed when it comes to the sitting position, so more extensive missing parts of the mesh surface have to be processed. The handheld scanner from *Artec* was first used to understand better which areas are problematic for the camera to capture, what choice of distance and angle is optimal for the obtention of scanner data and what degrees-of-freedom the body has in adopting positions which improve scan data-collection. A second proposed scanning procedure was carried out using both the handheld scanner from *Artec* and the body-scanner from *Vialux*. The entire procedure was performed using two healthy subjects because it was not possible or recommended to use a disabled person.

Scanning procedure for the sitting position using the Artec MHT handheld scanner

The starting point of the procedure for scanning the sitting position was to test the usability and efficiency of the *Artec MHT* handheld scanner in capturing data from such a problematic body posture. The area of interest for the scanning was the lower part of the body, which is the most challenging area of garment-construction for wheelchair-users. The tested person has to sit on a standard chair, without a seat-back; the feet placed flat on the ground, with knees bent. The arms have to be bent upwards at the elbow so as not to appear in the scan data for the lower part of the body. The person has to maintain the same position throughout multiple scan trials, so the feet have to be kept within marks drawn on the ground and the buttocks positioned on the chair in the same place each time (Fig. 3.12).

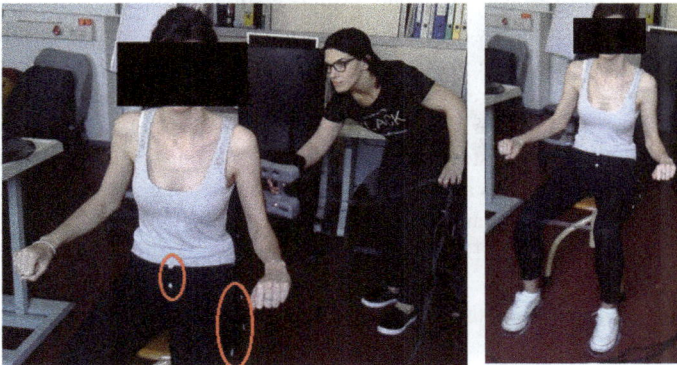

Figure 3.12. Scanning procedure using the MHT handheld scanner from Artec

The subject has to wear tight-fitting clothing in order to reflect the true shape of the body. For the purposes of later data processing it is necessary to affix white clay markers to various points of the body (Fig. 3.12). The location of a marker has to ensure maximum visibility to the editing program, so that it can identify them and merge the data. The markers divide the body in half so as to permit good delineation by the editing program. Multiple scans of the chosen position were performed, starting from different sides and various angles, in order to determine the best procedure for obtaining optimal data. The multiplicity of trials revealed that it is difficult to scan a large area of the body, using a handheld device, without suffering overlapping images or double mesh layers which can hinder the obtention of realistic images and correct data. A single leg was also scanned in order to obtain better data for the analysis of missing parts in the sitting position. The scan data were edited using *ArtecStudio* and the use of its processing tools to prepare a final 3D model involved [160]:

> ➢ Revising and editing the data – in this step, a sequence of frames obtained during the scanning procedure is saved as a single separate scan. The unwanted frames are removed, the misaligned areas (if any) are divided into separate scans and unwanted objects are removed from the scene. The final scan can be edited using the *Editor* icon-bar tool, where the coordinate plane can be established, the object can be rotated or moved and the scan surface can be processed by erasing or smoothing the surface.
>
> ➢ Alignment of scans – in this step, multiple scans are combined into a single model in order to obtain information concerning the relative position of each scan. The scan data must be converted into a single coordinate system: the alignment of the scans has to be done by dragging the scans together, with semi-automatic alignment or the creation of point pairs between the desired scans.
>
> ➢ After alignment, data registration is necessary in order to proceed to the next step – global registration. The registration algorithm converts all of the frame surfaces into a single coordinate system, using each pair of matching positions on the surfaces.
>
> ➢ The fusion of data into a single 3D model creates a polygonal 3D model by melting and solidifying the frames altogether.
>
> ➢ Final editing of the 3D model – in this step, the resultant fusion model can be processed by repairing, smoothing or simplifying triangulation errors in the model. The filling of holes or the erasure of unwanted parts of the object can also be carried out.
>
> ➢ Texture mapping – in this step, textures from the individual frames are projected onto the fused mesh.

Multiple scans were performed on the right leg, as an example, two of them were further processed in order to obtain optimum final scan data (Fig. 3.13). Unwanted frames were removed from both scans and then, using the *Editor* tool, outliers and poorly-scanned regions were deleted. The *Fine registration* algorithm was used to improve the scan surface. *Fine registration* is an automatically initiated algorithm that captures the frames of the scan precisely. The processed scan data (Fig. 3.14) were then used for the alignment procedure.

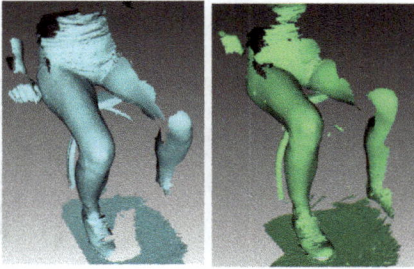

Figure 3.13. Various scan data for the right leg

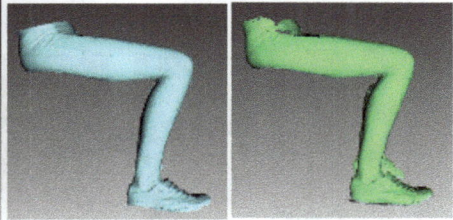

Figure 3.14. Scan data after revising and editing

A process was required in order to match pairs of points in the two scans (Fig. 3.15). White clay markers on the legs of scanned subjects were used to identify the same point in both scans. It was easy in that way to identify the matching points exactly and thus merge the two scans (Fig. 3.16).

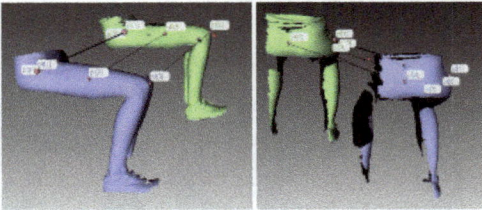

Figure 3.15. Creation of point-pairs between the two scans

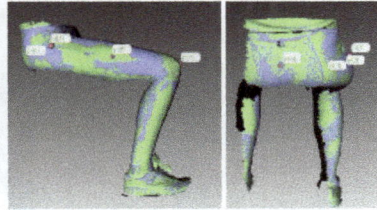

Figure 3.16. Alignment result

When alignment of the scans is complete and *Global registration* has been applied in order to obtain a single data-set within a common coordinate system, *Fusion* of all of the scan frames can be performed (Fig. 3.17). The result is a polygonal model that can be saved as an *.obj* file for further mesh-editing in *GeomagicStudio* (Fig. 3.18). *Geomagic* offers excellent and user-friendly solutions for mesh processing, so the repair and completion of missing areas of scan data could thus be finalized. Problem areas in the scanning procedure for the sitting position are the buttocks, the back of the thigh, the crotch and the inside of the leg; all places that light from the scanner cannot reach.

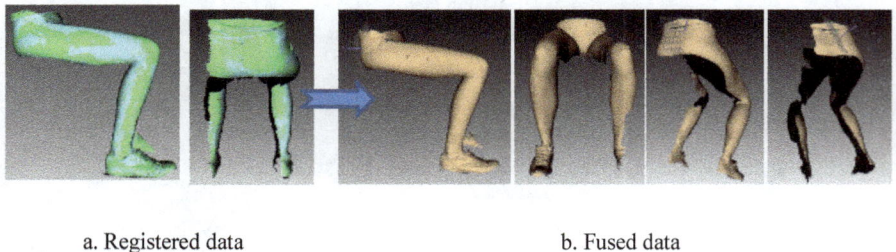

         a. Registered data   b. Fused data

Figure 3.17. Final data following Global registration

Mesh-editing in *GeomagicStudio* was performed according to the previously described scheme: filling-in of missing holes, smoothing the mesh surface and erasing unnecessary parts. Following mesh-processing, the leg was mirrored so as to create a complete lower part of the body that could then be used in 3D animation or simulation programs for virtual garment construction (Fig. 3.19).

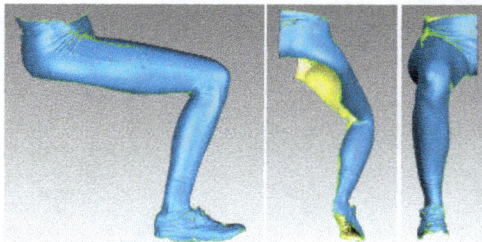

Figure 3.18. Right leg in polygonal phase in   *Figure 3.19. Polygonal mesh of the leg*

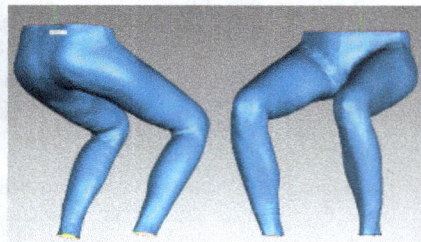

GeomagicStudio   *processed using GeomagicStudio*

Scanning procedure using Artec MHT and ViALUX zSnapper®

Having gained experience in using the scanning procedure for a normal chair, and knowing the parts which are difficult to scan, the next task was to develop a special chair (Fig. 3.20). The resultant article was made from transparent Plexiglas® which allowed the scanner beams to reach the thighs and underside of the buttocks. Only a tiny remaining area at the center of chair-seat, made from thicker Plexiglas® in order to receive the leg-leg, blocked the light from the scanner. The new chair also had a seat-back which was useful in supporting a paraplegic during scanning.

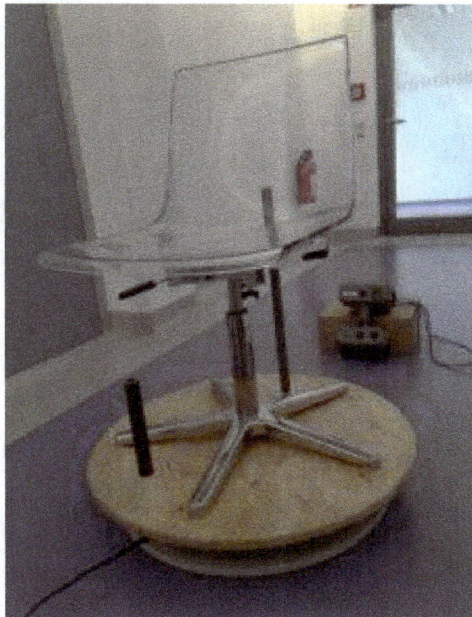

Figure 3.20: Plexiglas chair

The steps of the scanning procedure are detailed in Fig. 3.21.

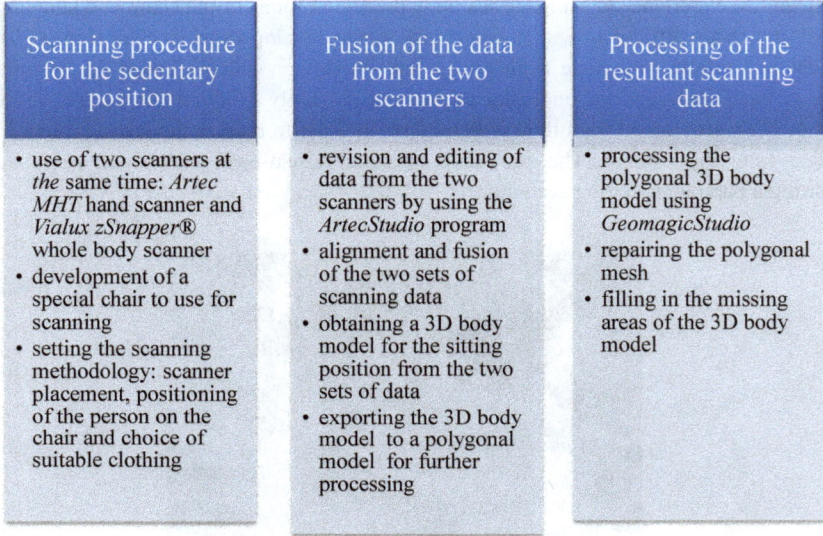

Scanning procedure for the sedentary position	Fusion of the data from the two scanners	Processing of the resultant scanning data
• use of two scanners at *the* same time: *Artec MHT* hand scanner and *Vialux zSnapper®* whole body scanner • development of a special chair to use for scanning • setting the scanning methodology: scanner placement, positioning of the person on the chair and choice of suitable clothing	• revision and editing of data from the two scanners by using the *ArtecStudio* program • alignment and fusion of the two sets of scanning data • obtaining a 3D body model for the sitting position from the two sets of data • exporting the 3D body model to a polygonal model for further processing	• processing the polygonal 3D body model using *GeomagicStudio* • repairing the polygonal mesh • filling in the missing areas of the 3D body model

Figure 3.21. Scan-data processing

Knowing that scanning the entire body using the *zSnapper®* device takes less than one minute, the newly designed chair was fixed to a circular stand that could be placed on the turntable of the *Vialux* equipment. This stand also has a guide-post which allows the maintenance of a fixed position when the wheelchair-user moves from one chair to another, and another guide-post which allows the height of the chair to be adjusted according to the height of the person (both of these posts are visible in Fig. 3.20). Scanning was performed by using the whole-body scanner and the handheld-scanner simultaneously in order to capture data from the buttocks area. The subject was required to sit on the chair with feet planted on the turntable, leaning against the seat-back and with the hands held at the back of the neck. The body-scanner was positioned so as to capture data from the rotating chair while the handheld-scanner was placed on the ground, 50cm from the rear of the turntable, with its field-of-view covering the buttocks (Fig. 3.22).

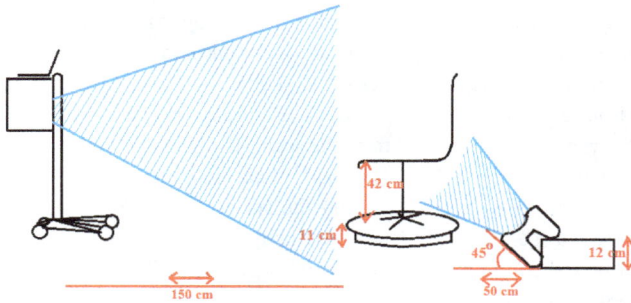

Figure 3.22. Scheme for scanner positioning in the proposed scanning procedure

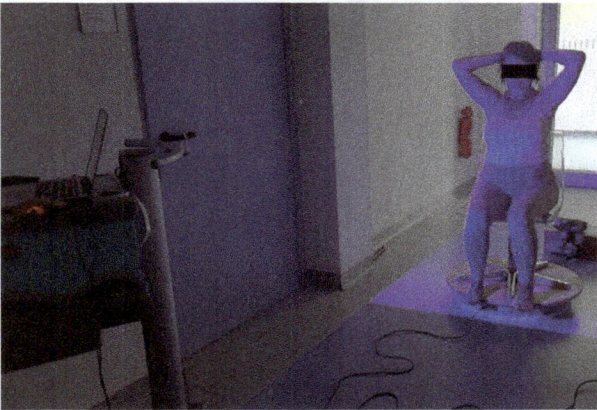

Figure 3.23. Scanning procedure using the Artec MHT and the ViALUX BodyLux®. zSnapper®

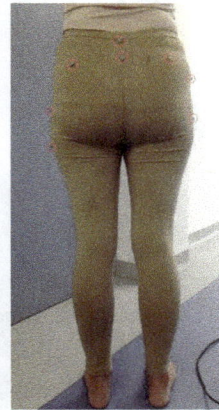

Figure 3.24. Clay markers

Both scanners were turned on simultaneously: the *ViALUX* scanner capturing data from the entire body and the *Artec* scanner covering just the lower part of the body (Fig. 3.23). The scanned person had to wear light-colored clothes, and clay markers were affixed to the body with the objective of later being able to merge the data from the two scanners (Fig. 3.24). The scan data which were obtained were further processed using *ArtecStudio* and *GeomagicStudio*.

Data fusion and surface reconstruction

The *ViALUX* body-scanner proved to have difficulties in capturing data from areas which the light beams could not reach directly. There were parts missing from the top of the thighs, the backs of the legs and the buttocks (Fig. 3.25). The *Artec MHT* scanner was used to capture data from the buttock-side (Fig. 3.26). They were processed using *ArtecStudio*. Any unnecessary frames were removed from the scans, e.g., those showing the chair guide-post and the area where the thicker Plexiglas® plate prevented the scanner light from reaching the body. After revising and editing the surface using *Editor* tools, two scans which comprised good (i.e. relevant) data were finally selected that could be merged so as to obtain a final model of the buttocks area (Fig. 3.27). The *Fine registration* algorithm was also used to improve the scan surface.

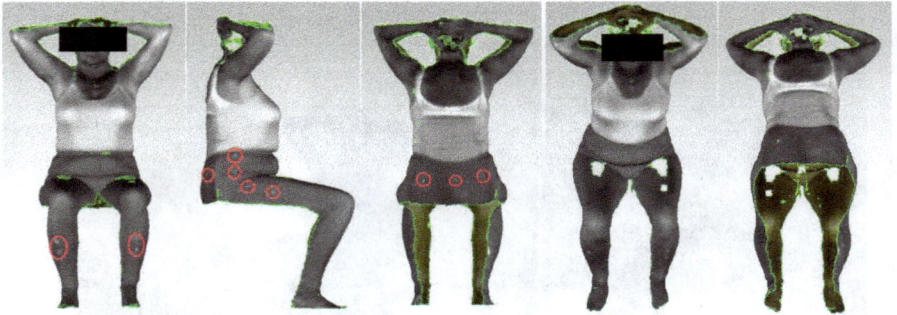

Figure 3.25. Scanning data obtained using the ViALUX BodyLux®. zSnapper®

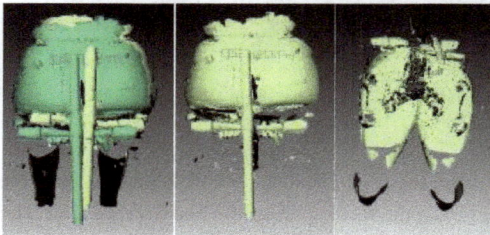

Figure 3.26. Scan data obtained using the Artec MHT

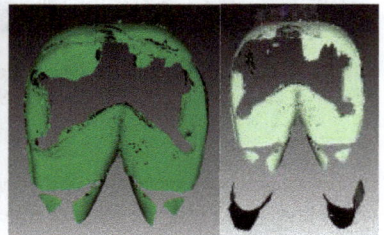

Figure 3.27. Scan data following revision and editing

The two scans were aligned by pairing points between them. The paired points were the locations where clay markers had been affixed prior to scanning (Fig. 3.28). By using the *Alignment* results (Fig. 3.29) *Fusion* of all of the frames was performed so as to obtain a single 3D model (Fig. 3.30). Although the 3D model of the lower part of the body did not include enough data from the legs area, the scanning method proved to be efficient in obtaining data on the thighs, buttocks and lower back. The data which were obtained using the *ViALUX* scanner, and those which were obtained for the lower parts of the body by using the *Artec* handheld scanner, could be fused so as to obtain a complete body image for the sedentary position.

Figure 3.28. Establishment of point-pairs between the two scans   *Figure 3.29. Alignment result*   *Figure 3.30. Fusion data*

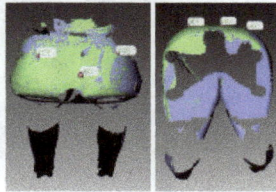

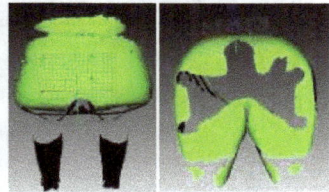

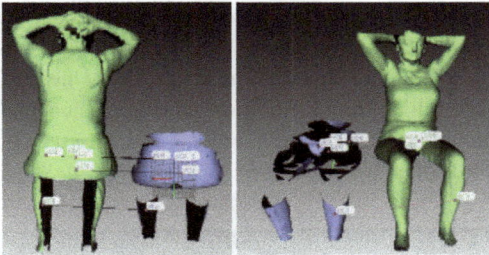

Figure 3.31. Establishment of point-pairs between scans obtained using Artec and ViALUX

Fusion of the data from the two scanners was performed in the following way. The resultant 3D model of the lower part of the body was stored in *ArtecStudio* and the *.obj* file of data from the *ViALUX* scanner was imported. The process involved the same steps as those for normal fusion: points in the two scans were first paired (Fig. 3.31), and the *Alignment* process was then applied (Fig. 3.32). The white clay markers on the legs of the scanned subject were used to establish the exact locations of the points in the two sets of data.

Precise overlaying of the same points in the two data-sets was therefore possible so as to obtain an accurate 3D body shape. The results of the alignment showed that the scanning protocol is accurate and precise when capturing data by using two different scanners at the same time. It is essential to calibrate both devices very carefully and to ensure identical positioning of the scanned body in order to guarantee error-free fusion of the data which is captured by the two different scanners. Both scanners were turned on simultaneously and synchronized so that exactly the same body position was scanned by both devices.

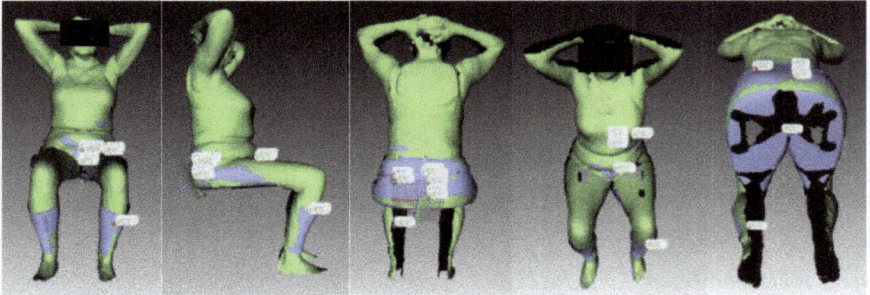

Figure 3.32. Alignment results

Following use of the *Fine registration* and *Global registration* processes in *ArtecStudio*, fusion of the two scans resulted in a new 3D body-model for the sitting position that could be further processed so as to fill in missing areas and improve the mesh surface (Fig. 3.33). The large amount of data resulting from fusion resulted in a 3D model having a polygonal surface that had to be simplified and reduced in order to make it useable by 3D programs. An *.obj* file was saved for further mesh-processing using *GeomagicStudio.*

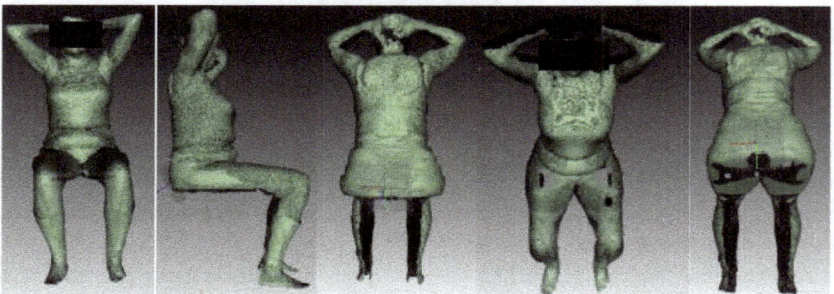

Figure 3.33. Fusion of the Artec and ViALUX scan data

The polygonal model imported from *ArtecStudio* (Fig. 3.34) exhibits many irregularities in the mesh surface because of the large amount of data that the program had to process. The simplest and fastest way to simplify the polygonal mesh is to convert the *Polygon object* into a *Point object* (Fig. 3.35). *GeomagicStudio* permits this via use of the *Convert to points* tool which removes the triangles while retaining the underlying points of the surface. The *Uniform* tool reduces the number of points to a specified density in the workplace for the point phase. The *Reduce noise* tool was then used to compensate for the scanning errors by moving the points to statistically correct locations so as to yield a smoother arrangement of the points (Fig. 3.36). When point reduction and improvement was complete the data could be converted, from a point cloud back to a polygon mesh, by using the *Wrap* tool. The result is a new 3D model, with fewer triangles, that further facilitates refinement of the polygonal mesh (Fig. 3.37).

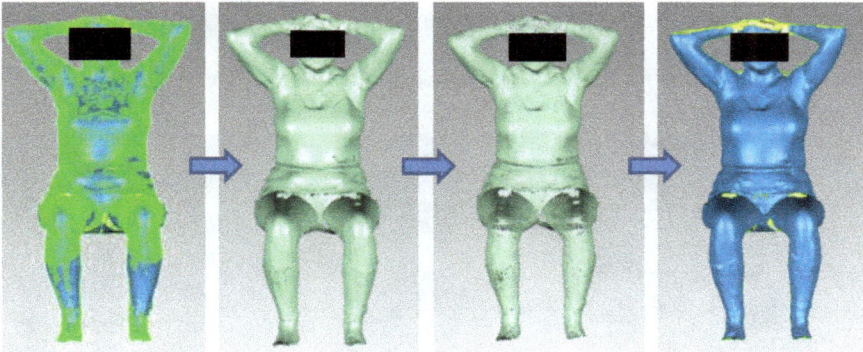

| Figure 3.34. 3D model in the sitting position in the polygon phase (4.134.639 triangles) | Figure 3.35. 3D model in the sitting position in the point phase (2.335.832 points) | Figure 3.36. 3D model in the sitting position in the point phase after processing (51.842 points) | Figure 3.37. 3D model in the sitting position in the polygon phase after processing (102.836 triangles) |

The merging of two sets of scan data which had been obtained by using two different scanners proved to be a feasible and valuable way of obtaining a 3D body model in the sedentary position, with fewer areas missing from the buttocks and the backs of the thighs (Figure 3.38). With the aid of these newly available scanned areas, the buttocks can be more easily repaired so as to obtain a realistic body-shape for the sitting posture. The mesh-editing process of *GeomagicStudio* was performed as described previously; filling-in the missing holes and smoothing the mesh surface using the *Remove spikes › Reduce noise › Relax* tools. Figure 3.39 shows the final 3D sedentary body-model in various side-views, with holes in the lower part of the body now filled and with a refined polygonal mesh.

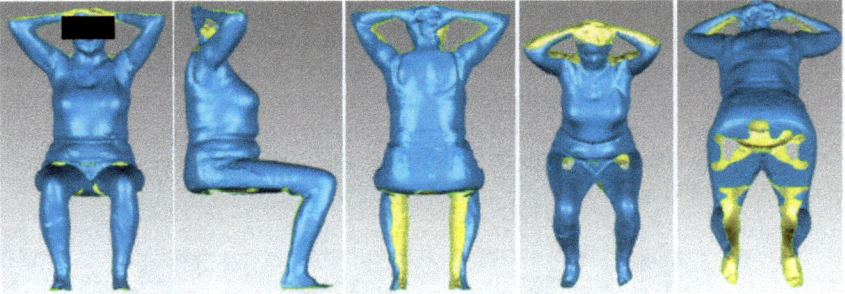

Figure 3.38. Views of the 3D sedentary body-model following triangle-reduction

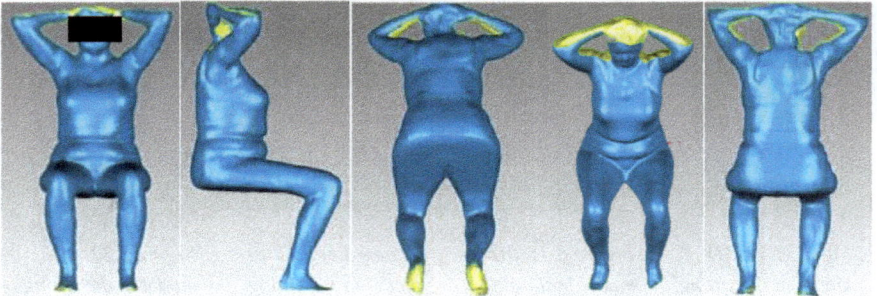

Figure 3.39. Views of the 3D sedentary body-model after filling-in missing areas and smoothing the polygon mesh

The scanning protocol which was established as a result of this study proved to be efficient in obtaining data and in minimizing the time required of the person being scanned. The first aim was to establish a method by which data from the lower part of a body in the sitting posture could be obtained. The second aim was to minimize the scanning-time so that a disabled person would not have to be kept in a particular position for too time. Although a first trial using the handheld scanner offered an overall view of what the capture of data for the sitting position involves, the method of capturing data by using two scanners proved to be an essential innovation in developing the best possible procedure to be adopted in this case. Having the newly-developed special chair, and knowing from previous trials the best technique for obtaining data within a short time, it was possible to validate the scanning trial by using a wheelchair-user to test the practicality of the chair and the possibility of scanning a paraplegic person by using the proposed method.

3.1.4 Scanning procedure for a wheelchair-user

A paraplegic person who had been using a wheelchair for 15 years was chosen for the next step of the study. He could move his trunk and arms, being immobile only in the lower part of his body. The lack of activity in the legs had led to a loss of muscle tone, leading in turn to a visible disproportionality between the abdominal region and the lower part of the body; his legs being thin while the upper part of the body was normal. This is a typical body structure for those who are disabled from the waist down.

It should be mentioned that the paraplegic person was informed about all of the steps involved in the scanning process. He filled in a signed agreement, as well as a questionnaire regarding the clothing problems that he habitually encountered.

The scanning procedure

The first step was to see whether a wheelchair-user could maneuver himself from his chair to the turntable. This did not prove to be a problem provided that the chair was stable; the subject could then move himself quickly and easily, without aid.

The second step was to see whether a paraplegic person could maintain the required posture for the duration of the scanning period: that is, to sit leaning against the chair-back while keeping the hands at the back of the neck. Having the chair-back with which to support himself, maintaining the position for less than one minute did not cause him any discomfort. Because of his inability to position his legs properly during the scanning process, it was necessary to use tape to keep them fixed in place (Fig. 3.40). The subject had to wear tight-fitting and light-colored clothes in order to observe the true body-shape during scanning.

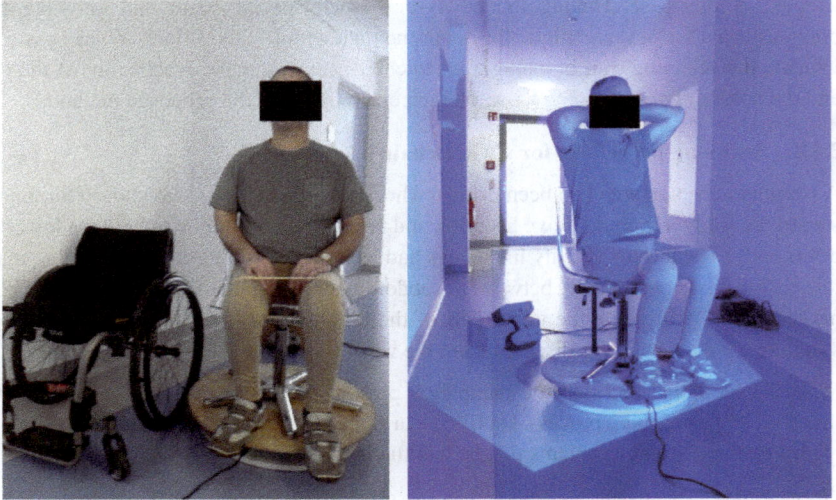

Figure 3.40. Scanning procedure for a wheelchair-user

Data fusion and surface reconstruction

The scan data which were obtained by using the *ViALUX BodyLux®. zSnapper®* (Fig. 3.41), and those which were obtained by using the *Artec* device (Fig. 3.42.), were merged by using the previously described procedure. The *Alignment* process in *ArtecStudio* was performed after locating the paired points (Fig. 3.43). The *Fine registration* and *Global registration* processes were then applied, and a complete 3D model was obtained (Fig. 3.44). Mesh-processing of the new model was pursued using *GeomagicStudio*. The elimination of triangles from the polygon mesh transformed the mesh surface of the 3D model into a stable surface that could be smoothed and refined.

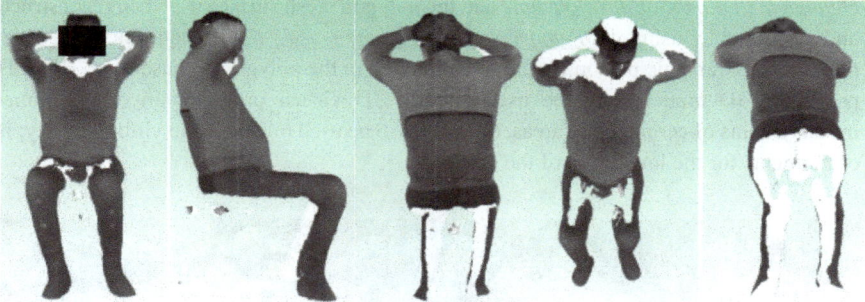

Figure 3.41. Scan data for the wheelchair-user, obtained using the ViALUX BodyLux®. zSnapper® scanner

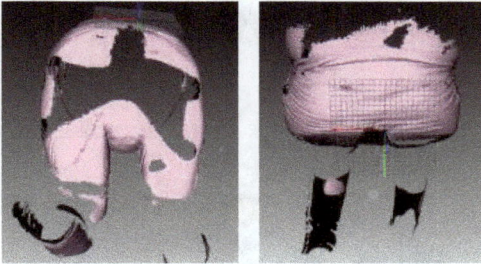

Figure 3.42. Fusion data obtained using the Artec MHT scanner

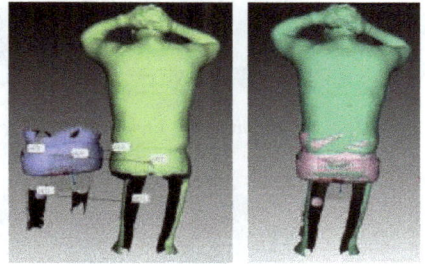

Figure 3.43. Creation of paired points, and alignment result

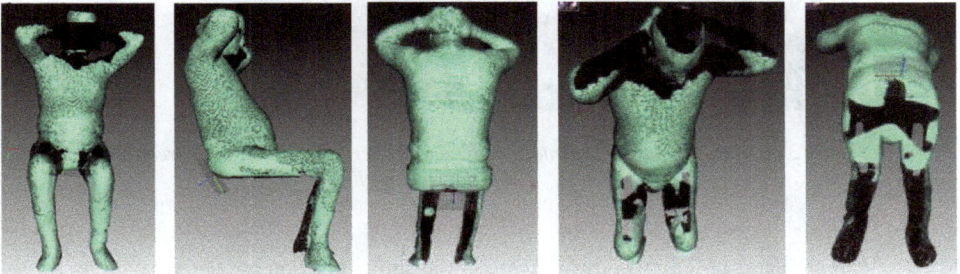

Figure 3.44. Fusion of Artec and ViALUX scan data

The areas which were missing from the scan data (Fig. 3.45) were filled in by using the *GeomagicStudio* tool *Fill single hole* and the polygon mesh surface was then reconstructed and smoothed by using the *Remove spikes › Reduce noise › Relax* sequence. A 3D body-model of a paraplegic person was finally obtained in the polygonal phase (Fig. 3.46). The polygonal 3D model could be used later in 3D virtual programs in order to make measurements of certain body areas, or could also be used to enable the virtual prototyping of garments for the lower part of the body.

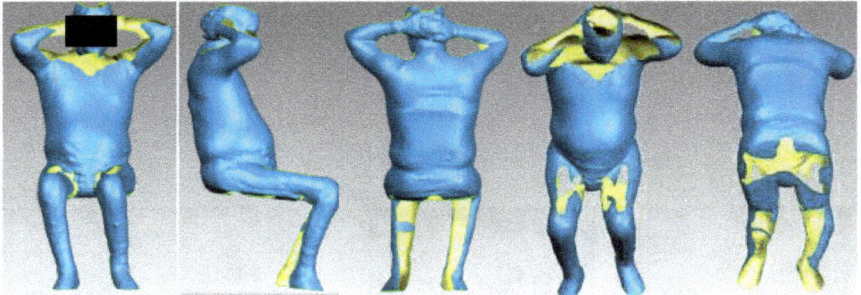

Figure 3.45. Views of the 3D sedentary body-model with missing areas in GeomagicStudio

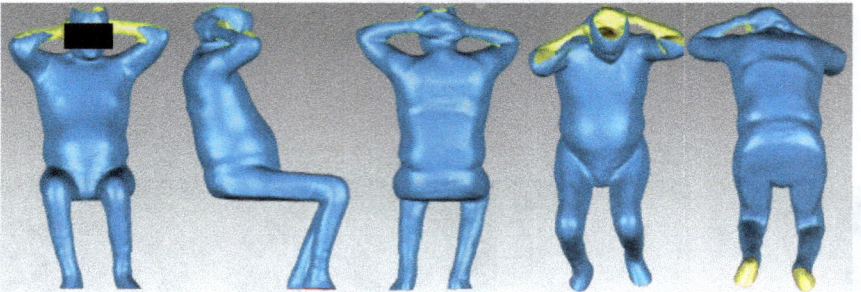

Figure 3.46. Views of the 3D sedentary body-model after filling-in missing areas and smoothing the polygon mesh

The problems posed by body-scanning in the sitting position are not yet entirely solved, but the concepts of a transparent chair and the use of multiple scanners were an essential first step towards gathering scan data more efficiently. Further research can be envisaged with regard to improving the thicker Plexiglas® area of the chair and reducing the amount of scan data missing from the buttocks area.

3.2 Body-posture animation using a kinematic template model

The objective of the virtual body animation was to explore the possibility of obtaining a sedentary 3D body-model for use in the virtual development of garments. The virtual prototyping of garments today offers the opportunity to design comfortable and functional clothing. Most of the CAD systems for pattern design and fit simulation use however virtual body-models having a standard shape and an upright posture. One of the objectives of the study was therefore to obtain a realistic virtual body-model in the sitting position. This led on to the use of 3D animation programs to deduce scan data, for other postures, from the standard posture. Various positions of the lower part of the body, going from the standing position, passing through the various intermediate positions and ending with the sitting position, were studied. In order to obtain an animatable "kinematic body model", scan data for a subject (Fig. 3.47) were merged with a "kinematic template model"[1] which had been developed during a previous study [97] by using the *3dsMax* program (Figure 3.48). The *3dsMax* program comprises 3D modeling and rendering software that helps to create virtual scenes for design visualization and computer games [106].

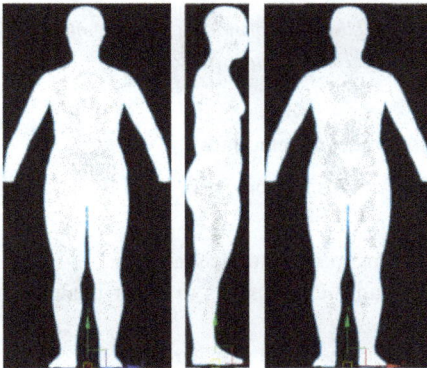

Figure 3.47. Female scan images   Figure 3.48. Kinematic templates

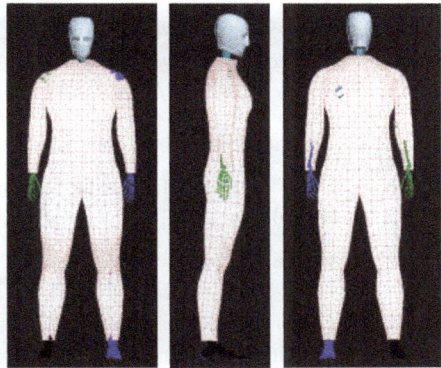

---

[1]the template represents the designed kinematic human model in 3dsMax consisting of a skeleton, muscle system and surface mesh

An *anatomically-based animation* for the *3dsMax* program has been presented [97] which uses the *Skeleton-driven Deformation* technique. This means that the template is composed of a skeleton that is connected to a skin surface. The kinematic template in this case consists of a skeleton, a surface mesh and a muscle system. The musculoskeletal template was designed using basic human anatomical principles.

The template in question was intended to treat only the lower part of the body; the entire system of bones, muscle and skin being available for use in various animated positions (Figs. 3.49 - 3.50) [97]. The kinematic template was merged, by using the *3dsMax* program, with scan data by adapting the template surface to the surface of the target scan model. By merging the kinematic template with the scan data, it was possible to animate the scanned surface and obtain various postures of the lower part of the body.

Figure 3.49. Representation of the musculoskeletal template in various positions [97]

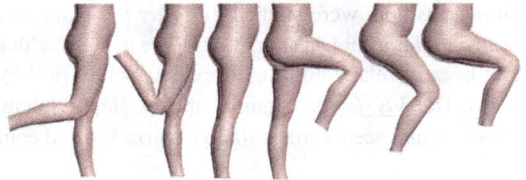

Figure 3.50. Skin deformation of the template model in various positions [97]

3.2.1 Adaptation of the kinematic template model to scan data

In order to obtain animatable scan data by using the *3dsMax* program, the following steps have to be followed [102]:

➤ manual interactive adjustment of the template posture to the scan data,
➤ width- and length-scaling of the bones and muscles so as to adapt them to the proportions of the scanned object,
➤ skinning - the adaptation of the template vertices to the scan surface-data,
➤ refinement of the new surface net, and
➤ automatic transfer and calculation of the interpolated weights.

As well as posture adjustment, which is performed manually, the other steps are carried out by using various scripts from *MAXscript*, which is a programming language that is integrated into the *3dsMax* software. All of the scripts used have been described in detail elsewhere [97]. Manual adaptation of the template to the scan data is done by dragging them together (the crotch areas of both data-sets are in the same position). Both of them have to be proportionally located in the same place: in Fig. 3.51 white indicates body-shape scan data and the skeletal model while red indicates the kinematic template. Further scaling of the bones and muscles, in length and width, was carried out by applying the *.scaleMuscleDynamics* and *.scalePelvisWidth* files of *MAXscript*. The latter program permits the changing of scaling factors for the right and left leg or thigh, torso or pelvis. The scaling factor may vary from one value to another, depending upon the height or width of the person from whom the scan data were obtained. It is essential to adjust the scaling exactly to the dimensions of the scan data, the kinematic body will otherwise have dimensions which differ from those of the scan. It was necessary to adapt manually the leg position of the template to one of the scans (Fig. 3.52). No adaptation was needed for the arms because a musculoskeletal template had not yet been developed for the upper part of the body and was thus irrelevant to this study. The two adjusted data-sets could be merged in the next step by adapting the template vertices to the scan surface in order to obtain a kinematic body-model.

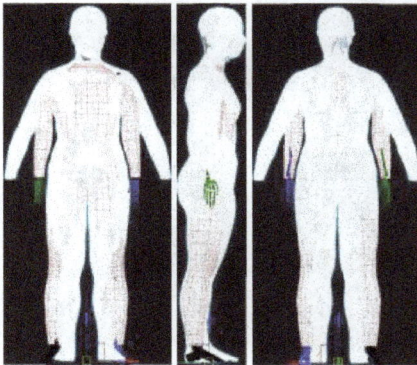

Figure 3.51. Adaptation of the template to the scanned body

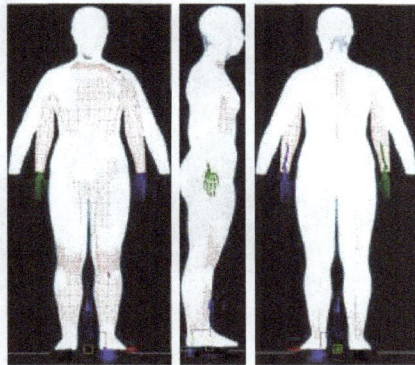

Figure 3.52. Adaptation of the template to the scanned body, following leg adjustment and muscle scaling

The skinning procedure was carried out by adapting the template vertices to the scan surface data. The *Convert to editable Mesh* tool was used to work with the mesh template (Fig. 3.53). The *adapt TemplateSkin* script connected the template vertices and the scan surface, thus generating a new kinematic body-model having the shape characteristics of the lower part of the body as deduced from the scan data (Fig. 3.54). The various colorings of the kinematic body-model areas indicate the different scaling calculations which have to be applied to the muscle modifications during animation.

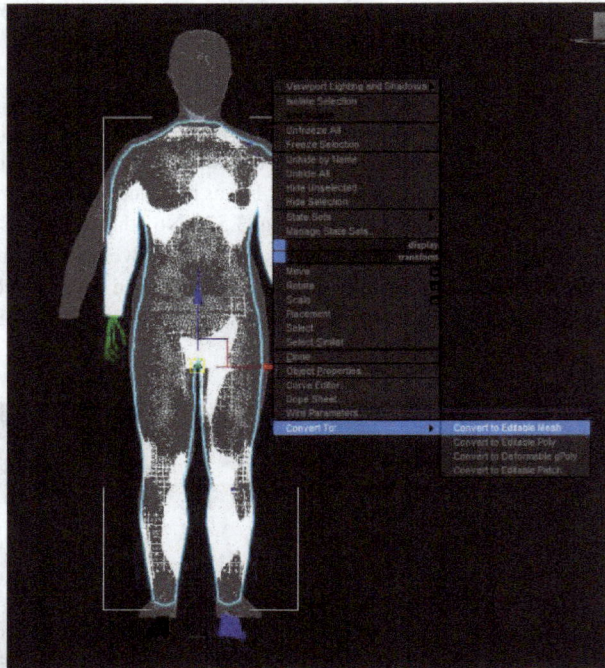

Figure 3.53. Conversion-to-editable-mesh procedure

Each area requires different scaling algorithms to be applied to the muscles and mesh area in order to be able to model realistic deformations during motion. The obtained mesh surface tends to be divided into relatively large rectangular areas, and this could impair realistic mesh-deformation during body animation. Refinement of the mesh surface was therefore performed by applying the script sequence: *.divideTemplate* › *.adaptTemplateSkin* › *.addSkin*. The result was a kinematic body-model having an improved mesh that could follow more precisely the shape of a real body. The final step was the automatic transfer of the weights, and calculation of the interpolated weights. The scripts, *.readWeightData* and *.calcAndSetDivideWeights*, were applied and the lower part of the kinematic body-model was finally ready to be animated (Fig. 3.55).

The entire procedure of adapting the kinematic template model to the scanned object was applied several times, with various scans covering different sizes and differing adjustment techniques. An optimal kinematic body-model could be obtained for future use in the animation process. The experience thus gained helped to develop faster and more reliable procedures for working with the program tools.

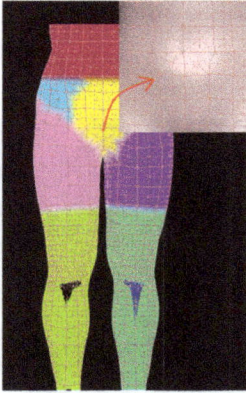

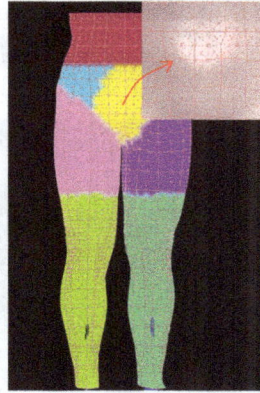

Figure 3.54. Kinematic body-model   *Figure 3.55. Kinematic body-model with refined mesh and applied weights*

3.2.2 Animation of the kinematic model to obtain various positions of the lower body

A realistic 3D human animation has to consider the deformation of the body during motion. The musculoskeletal template model which was developed, using the *3dsMax* program, for the lower part of the body was an essential part of the kinematic body-model animation. With the help of the skeleton and muscle systems, the kinematic body-model could take on realistic shapes during the animation. A sequence taken from the animation process can be seen in Fig. 3.56. From the sequence of positions occurring during dynamic motion, various postures with bending angles ranging from standing to sitting, were selected in order to study further the problem of garment construction for the sitting position. The lower parts of selected bent positions of the body were converted to *EditablePatch*, which offered a smoother geometry for the mesh edges and was exported as an *.obj* file for *3DDesignConcept* from *Lectra* (Fig. 3.57).

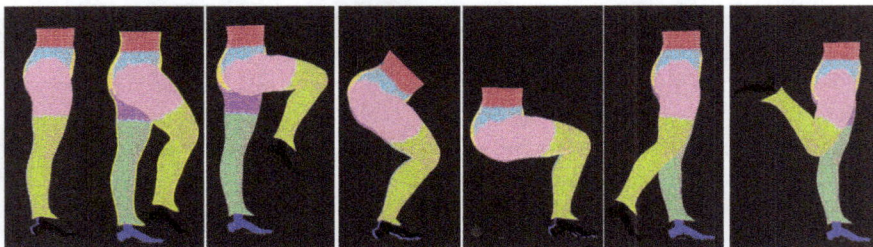

Figure 3.56. Motion sequence of the kinematic body-model

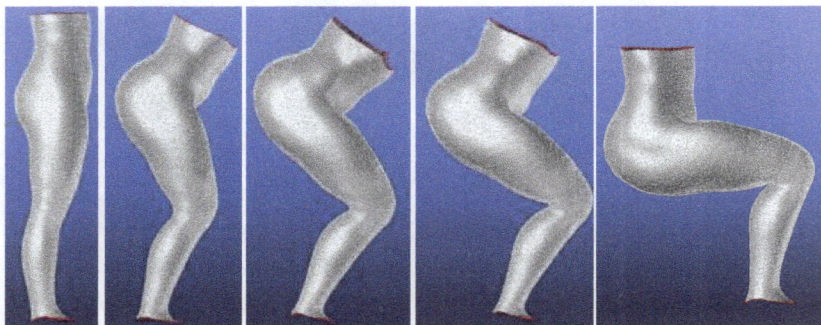

Figure 3.57: Sequence of positions between standing and sitting for the lower part of the body in 3D DesignConcept, [source Lectra]

The exported positions were further used to study the possibility of 3D construction of a pair of pants. It can be seen that for the sitting position there is no additional support for the sitting area that could offer a realistic body-shape in this particular case.

3.2.3 Body-posture simulation using scan data obtained using the ViALUX BodyLux®, zSnapper®

The feasibility of the kinematic body-model was demonstrated in this part of the study by changing the posture from standing to sitting. The idea arose from a previous scanning procedure for a healthy female in the standing and sitting positions. A healthy subject was again used to verify and validate the proposed animation method. The experience gained by using the *3dsMax* program facilitated the creation of a kinematic body-model from a scan of a standing body, obtained using the *ViALUX* scanner. The procedure followed the same steps as those described in section 3.2.1; it used the same kinematic template as that of the *3dsMax* program, together with *ViALUX* scan data (Fig. 3.58). The kinematic template was merged with the scan data for the female subject, and the scaling factors had to be adjusted to the size of the scan. The merged data (Fig. 3.59) were transformed into the kinematic body-model (Fig. 3.60); executing the skinning procedure, refining the new surface and applying weight transfer by using the necessary scripts.

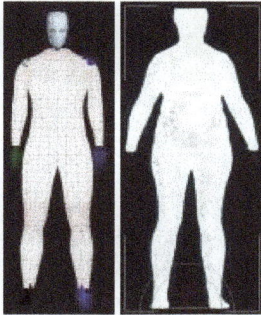

Figure 3.58. Kinematic template (left) and scan data from ViALUX scanner (right)

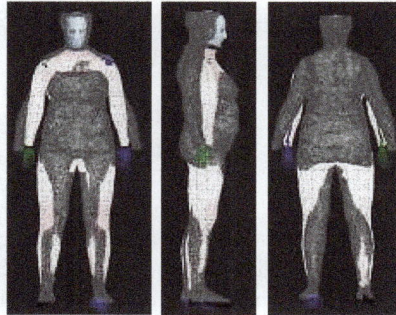

Figure 3.59. Adaptation of the template to scan data following leg adjustment and muscle scaling

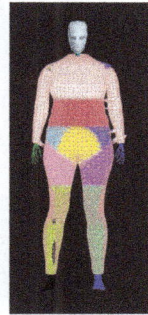

Figure 3.60. Kinematic body model with refined mesh and applied weights

The kinematic body-model was animated in order to demonstrate that, even with the atypical conformation of a tested person who has an abdominal area which is more prominent and legs which are thinner, the template of *3dsMAX* can adapt to the scan shape and dimensions. The motion sequence of the animation process maintained the same realistic shape of the lower part of the body (Fig. 3.61).

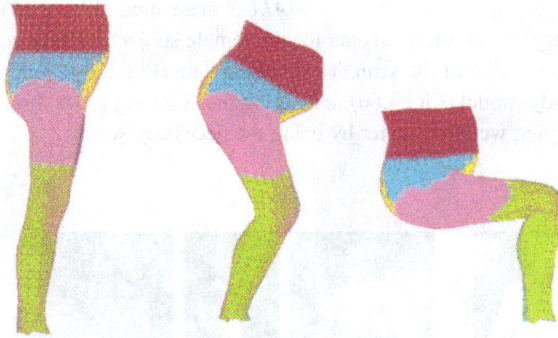

Figure 3.61. Motion sequence of the kinematic body-model

The standing, bent and sitting positions were selected during the animation process and converted to *EditablePatch* to give a smoother mesh area (Figs. 3.62 - 3.64). The *GeomagicQualify* program used the results to make a dimensional comparison between them and the scan data for those specific postures.

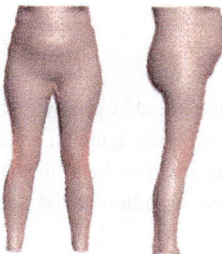

| *Figure 3.62. Views of the standing lower part of the body* | *Figure 3.63. Views of the bent lower part of the body* | *Figure 3.64. Views of the sitting lower part of the body* |

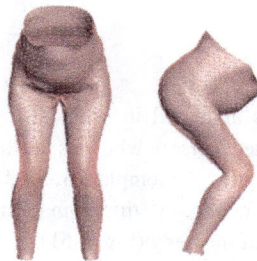

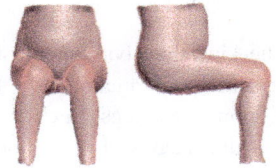

3.2.4   Comparison between kinematic body-model and scan data

This procedure tests the reliability of a kinematic body-model in exhibiting a realistic body shape during motion. Alignment of the two data-sets was performed using the *GeomagicQualify* program. The latter can make accurate comparisons of scan data or of scan data and *.iges* data files. The program also offers, together with tolerance and deviation computations, a color-mapping of results that can be easily understood from the displayed graphics [166]. Alignment of the scan data and kinematic body-model was performed by using *ManualRegistration*; defining corresponding points in overlapping areas of each data-set. The scan data were chosen to be the reference object following manual registration and the kinematic model was the tested object. The *BestFitAlignment* program was applied so that the tested object could be automatically moved so as to share the same physical space as the reference object. The resultant 3D alignment (Fig. 3.65) was further processed by applying *3DCompare*, which generates a three-dimensional color-coded mapping of the differences between the Test and Reference objects (Fig. 3.66).

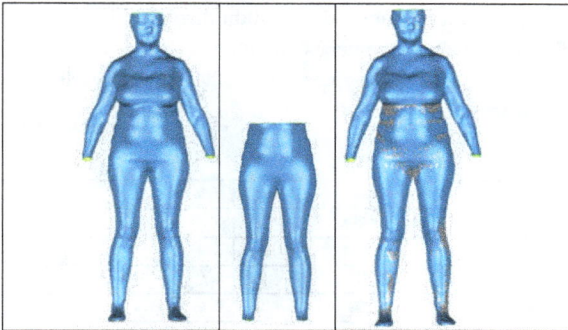

Figure 3.65. Scan data (left), kinematic body-model (center) and 3D alignment (right)

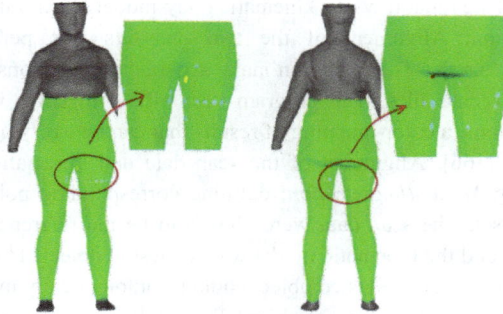

Figure 3.66. 3D color comparison

Upon analyzing the resultant deviation-distribution display (Fig. 3.67) it can be seen that approximately 99% of the test area matches the reference area, with a deviation of less than one millimeter (-0.850 to +0.850mm, green). The other 1% of overlapping points (yellow, blue and red) range from about ±3mm to ±17mm.

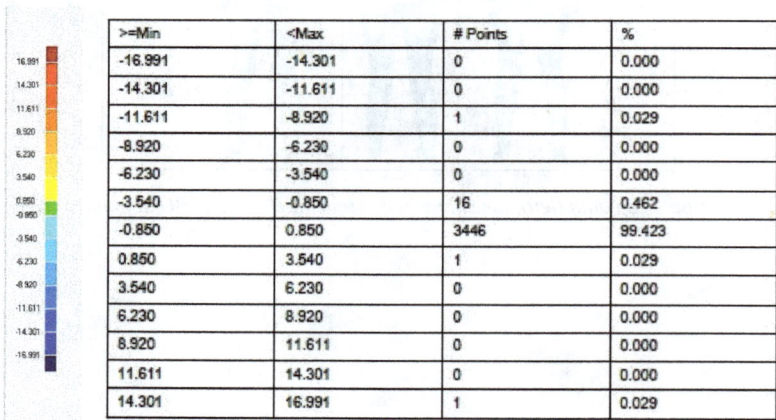

>=Min	<Max	# Points	%
-16.991	-14.301	0	0.000
-14.301	-11.611	0	0.000
-11.611	-8.920	1	0.029
-8.920	-6.230	0	0.000
-6.230	-3.540	0	0.000
-3.540	-0.850	16	0.462
-0.850	0.850	3446	99.423
0.850	3.540	1	0.029
3.540	6.230	0	0.000
6.230	8.920	0	0.000
8.920	11.611	0	0.000
11.611	14.301	0	0.000
14.301	16.991	1	0.029

Figure 3.67. Deviation distribution in [mm]

Materials Research Foundations **110** (2021)   https://doi.org/10.21741/9781644901557

The alignment between the two data-sets shows that the kinematic template of the *3dsMax* program is efficient in creating a kinematic body-model that can maintain the same body shape as that of the scan data. The subject was scanned in a bent but comfortable position in order to check the shape of the kinematic body-model for the bent posture. The resultant scan data and the lower part of the kinematic body-model (Fig. 3.69) were aligned using *GeomagicQualify* as previously described. Judging by the 3D alignment (Fig. 3.70) the kinematic body-model could also adopt a realistic shape for the bent position.

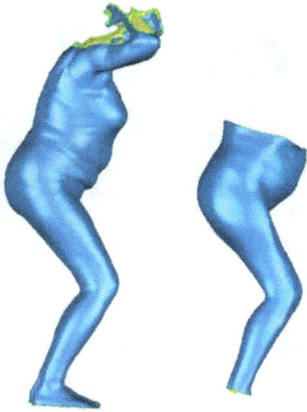

Figure 3.69. Scan data (left) and
kinematic body-model (right) for
the bent position

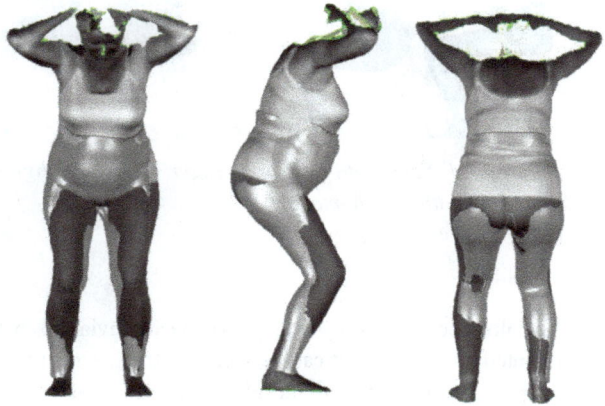

Figure 3.70. 3D alignment for the bent position

Scanning was carried out for a person sitting on the Plexiglas chair in order to align the results with the sitting posture of the kinematic body-model. This measured the distance between the knees and feet of the kinematic model, and the person had to adopt exactly the same position of the legs during scanning. Using the new scan data and the kinematic body-model in the sitting position (Fig. 3.71), 3D alignment between them was carried out.

Without even performing a *3DCompare* of the resultant alignment (Fig. 3.72) it could be seen that there was a big difference between the shape of the scanned body and that of the kinematic body. The shape of the kinematic body-model is bulkier in the buttocks and calf area than is the scan data. The explanation for this is that, during the animation process, the kinematic body does not have a chair to support itself, as a person does while sitting. The real calf-area muscles are more relaxed in the sitting position and the flesh in the buttocks area is flattened against the chair seat. The kinematic body could also not adapt itself to the natural shape of the waist area when in the sitting position.

Figure 3.71. Scan data (left) and kinematic body-model (right) in the sitting position

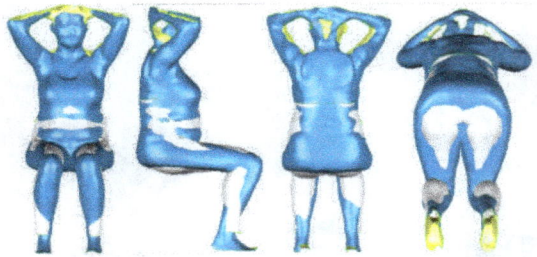

Figure 3.72. 3D alignment for the sitting position

The color-coded mapping (Fig. 3.73) reveals deviations between the scanned body and the animated body-model. It can be seen that the green surface of optimal overlapping points occupies a small area of the alignment, the rest being occupied with areas of more considerable deviation. Upon analyzing the deviation-distribution display (Fig. 3.74), it can be seen that approximately 48% of the surface area has a deviation of ±3.39mm, while 45% of the surface deviates by -24 to 35mm and the remaining 7% is off by ±56mm. It can be concluded from an analysis of the dimensional deviations that the animation process cannot offer proper dimensional changes to scan data for a body in the sitting position. The sitting position which is obtained via animation of a kinematic body-model is therefore not suitable for obtaining a good 3D sedentary body-model for use in a 3D garment-development environment.

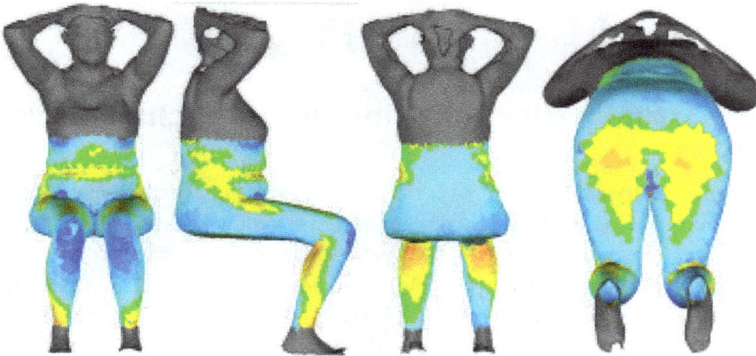

Figure 3.73. 3D color comparison

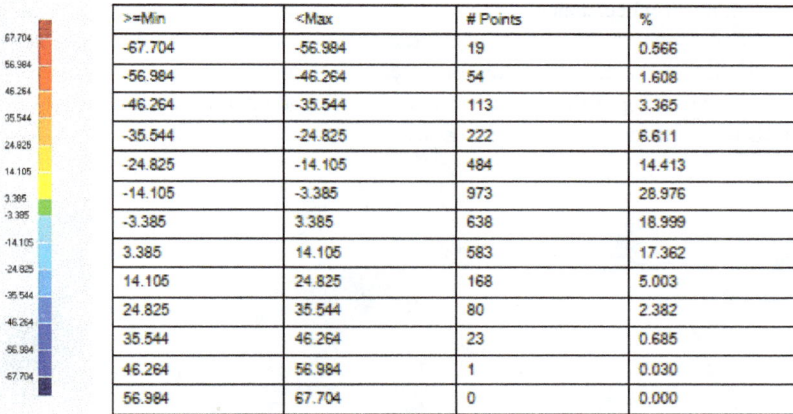

>=Min	<Max	# Points	%
-67.704	-56.984	19	0.566
-56.984	-46.264	54	1.608
-46.264	-35.544	113	3.365
-35.544	-24.825	222	6.611
-24.825	-14.105	484	14.413
-14.105	-3.385	973	28.976
-3.385	3.385	638	18.999
3.385	14.105	583	17.362
14.105	24.825	168	5.003
24.825	35.544	80	2.382
35.544	46.264	23	0.685
46.264	56.984	1	0.030
56.984	67.704	0	0.000

Figure 3.74. Deviation distribution in [mm]

Although animation of a kinematic body is generally an excellent way of observing body changes during motion, in the sitting position where other objects (e.g., a chair seat) are involved and have to be taken into account, animation does not generate a realistic shape for the body. More studies are required in order to solve the problem of obtaining a 3D body-model for the sitting position by using animation programs.

Chapter 4

4. Virtual pattern-making for wheelchair-users

The clothing industry has switched, over time, from conventional pattern-making on paper to CAD pattern-making. A clothing computer-design system will, as mentioned previously, comprise three integrated procedures: 2D pattern-design, 3D garment-construction and virtual clothing try-on simulation [167]. With the development of 3D CAD technologies the idea of designing a garment directly on a 3D body-model became an innovative solution to the construction of garments for people with disabilities. Few studies have analyzed 3D garment construction for people with scoliosis [145,171,172]. When it comes to the virtual prototyping of clothing for wheelchair-users, the current literature is based mainly upon the 2D-to-3D technique [89,91,94,95]. All of these studies approached the concept of garment-fitting to wheelchair-users by using CAD systems in experimental methods of pattern modification. The 3D-to-2D technique is described in the following chapter with the aim of finding a suitable solution for the construction of garments giving a good fit to the body in the sitting position. The main steps which are involved in the 2D-to-3D and 3D-to-2D processes are presented in Fig. 4.1.

3D-to-2D virtual prototyping	2D-to-3D virtual prototyping
The construction of tight-fitting pants for various body positions	The construction of a basic pants and fit simulation
Obtaining 2D patterns from the flattening procedure	Modification of the basic pants patterns using the 3D model dimensions
The construction of a tight-fitting pants model on a defined body position	Fit simulation of the modified basic pants
Obtaining 2D patterns from the flattening procedure	Designing a pair of pants for a male wheelchair-user
Fit simulation for the obtained tight-fitting pants model	Pants fitting by a wheelchair-user

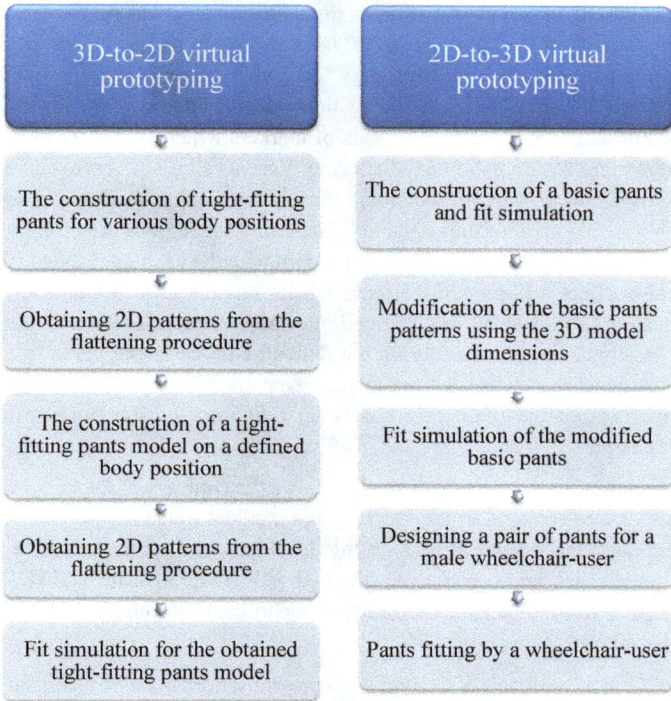

Figure 4.1. 3D-to-2D and 2D-to-3D prototyping

4.1   3D-to-2D virtual prototyping

The 3D-to-2D technique involves constructing the garment directly on a virtual body-model in the 3D environment, with the 2D pattern pieces then being generated by flattening the created regions. The construction-lines which define the shape of the garment are first drawn on the virtual body-model, and 3D patterns are then generated on the basis of the drawn regions. The 3D pattern pieces which are obtained are flattened into 2D patterns by applying a mesh process, using a mathematical algorithm, and the 2D patterns are then finally ready to be used for the creation of the garment model. The garment which is developed by using the 3D-to-2D technique follows the silhouette and the measurements which define the 3D body-model. Modification of a basic model over several steps, until it matches the silhouette of the person, is thereby eliminated and the time needed for product development is reduced.

For this purpose the 3D virtual prototyping of tight-fitting pants was performed by using *DesignConcept* from *Lectra*. This is one of *Lectra*'s solutions for textile design and development in the automotive industry, and for manufacturers which use functional fabrics [170]. *DesignConcept* is a product which was originally intended for car-seat manufacturers, but which also meets the needs of markets which specialize in functional clothing, from garment design up to manufacturing. Various styles of curves can be drawn on surfaces by using these program tools, and these curves and edges usually define the triangular meshing of the regions thus obtained. These meshed regions can be flattened and the 2D patterns can be associatively modified in accord with the 3D model style and curves. The program also offers a library of different fabrics having various characteristics, or allows other desired fabrics to be incorporated into the database. Realistic simulation of the fabric deformations which occur during the flattening process, and the tools available for treating manufacturing details such as notch or seam allowances, make *DesignConcept* a full-option program that can be used not only for functional textiles but also for tight-fitting clothing design.

A tight-fitting garment is a product that closely follows the contours of the body. The large demand for this type of clothing in the sporting and medical fields necessitates the study of divers functional requirements. Although the development of patterns on the basis of virtual body models has been intensively studied [133,134,139,140,141,174,175,176], fitting problems still need to be addressed further due to the multifunctional requirements placed on the garments.

A virtual model for the lower part of the body was previously obtained by using the *3dsMax* animation program, and scan data from a female subject (see Chapter 3.2.2.). A study which was performed by using *DesignConcept* aimed to work out which of the positions that had been obtained using *3dsMax* was ideal for designing a pair of tight-fitting pants; bearing in mind that the 3D model must be successfully flattened into 2D patterns. Many bending angles for the virtual body-model were checked in order to see which one was the most appropriate for designing the 3D pants model. The latter was designed by defining specific patterns for those areas of the body (knees, pelvis, thighs) where, in the bending position that a wheelchair-user typically adopts, modifications are needed.

The model has five patterns for the front and three for the back. The 3D pants-pattern outlines were drawn by using the *Draw curves* tool. The mesh regions for the patterns were obtained by applying *Create regions from curves,* and the flattening procedure was then applied in order to obtain the 2D patterns (Fig. 4.2). A mesh region is composed of *Links, Vertexes and Faces* (Fig. 4.3). There were differing *Link* lengths for the mesh regions, which aided in selecting the optimum dimensions for the flattening procedure. Three positions involving different bending angles are shown in Figs. 4.3 to 4.5.

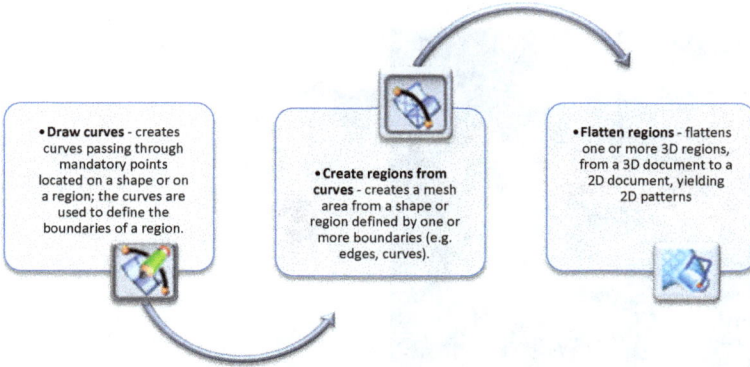

Figure 4.2. Automatic conversion of a 3D surface into 2D patterns

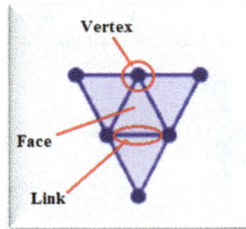

Figure 4.3. Sub-Division of a Mesh region

The first degree of bending which is selected corresponds to a sitting position with a 90° bend in the knee and trunk (Fig. 4.4a). It can be seen, following flattening, that a bending posture which involves extreme curvature in the buttocks and knee areas does not permit the creation of good patterns (Fig. 4.4b). The difference in mesh triangles in going from 3D to 2D is too large to produce realistic 2D patterns which have the same surface dimensions as those of the 3D document. The 2D patterns overlap in the groin area, and those for the back of the knee do not maintain a realistic shape.

Upon changing the degree of bending of the posture, as shown in Figs. 4.5 and 4.6, the patterns for the groin area are realistically flattened when going from 90° to 100°. The pattern for the back of the knee has the proper shape if the degree of bending is increased from 90° to 130°.

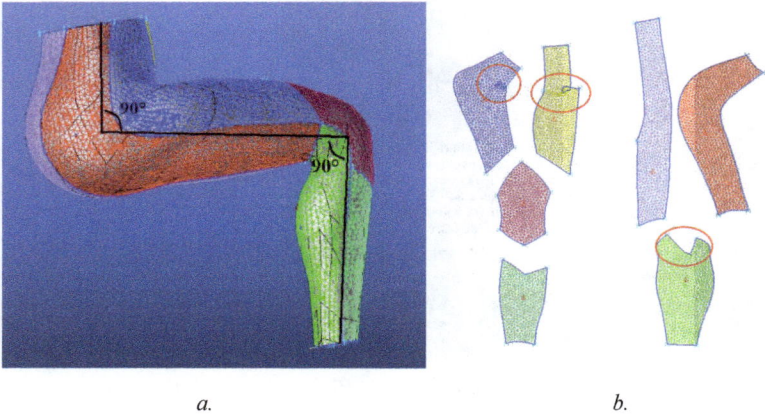

a.   b.

Figure 4.4. a - 3D pants model for a posture with 90° bending of the knee and trunk, and b - 2D patterns with flattening errors

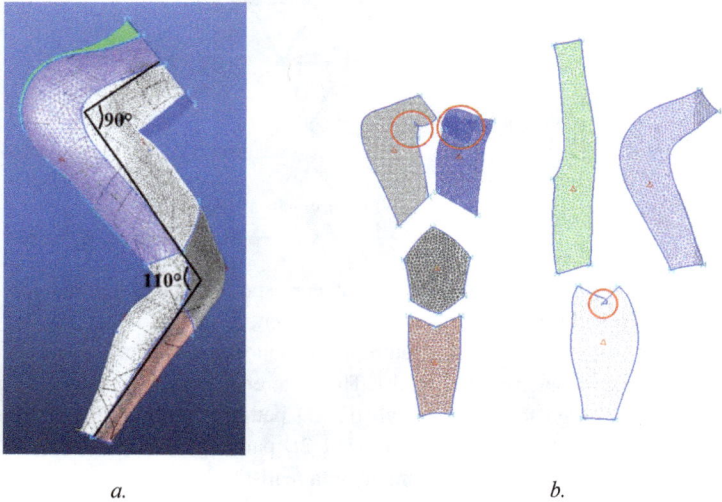

a.   b.

Figure 4.5. a - 3D pants model for a posture with 110° bending of the knee and 90° bending of the trunk, and b - 2D patterns with flattening errors

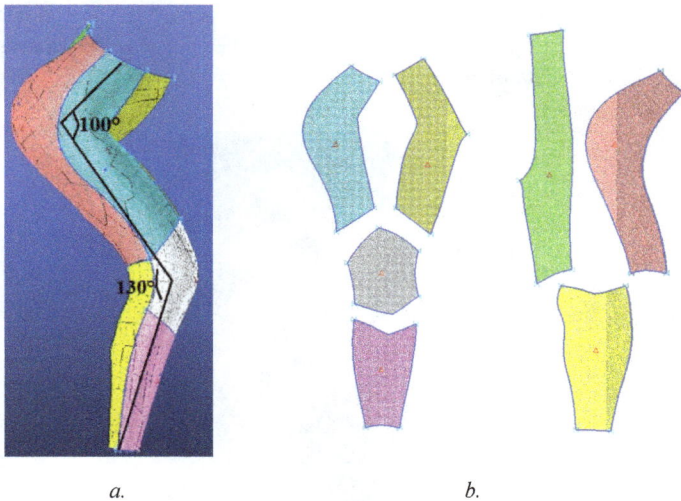

a.   *b.*

Figure 4.6. a - 3D pants model for a posture with 130° bending of the knee and 100° bending of the trunk, and b - 2D flattened patterns

4.1.1   The construction of tight-fitting pants on a defined body position

Tight-fitting pants were designed on a virtual body-model with a bending angle between the thighs and trunk of 90°, and a 110° bending of the knee (Fig. 4.7), taking into account the results of the pattern-development for various bending angles of the knee. In order to ensure optimum construction on the 3D body surface the polygonal mesh was transformed into a free-form shape by extracting the NURBS surface in *GeomagicStudio*. A NURBS (Non-uniform Rational B-spline) surface is a mathematical description which is commonly used for generating and representing curves, surfaces and volumes. It offers a unified mathematical basis for handling analytical and free-form shapes.

The NURBS model is used to recover information concerning a required spatial shape with great flexibility and precision, and it became a standard for CAD (Computer-Aided Design) systems due to its excellent mathematical, numerical and algorithmic properties [177,178].

Following 3D-2D processing of the chosen position (Fig. 4.3) the boundary curves for the model patterns were drawn (Fig. 4.8) and the mesh regions were then created (Fig. 4.9). It was necessary to create multiple regions for areas of high curvature so that the flattening procedure would be successful. A posture which was close to the sitting one was chosen in order to see whether it is possible to design the pants model. The groin-area patterns were divided into two parts in order to solve the overlap-problem. The optimum angle for the pattern covering the back of the knee was found to be 110°. The 2D patterns were obtained by using a *Link length* of 6mm and a *Vertex angle* of 135° (Fig. 4.10).

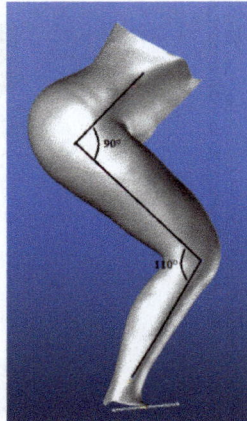

Figure 4.7: Bending posture for designing the tight-fitting pants model

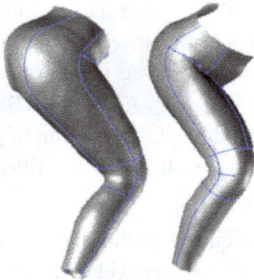

| *Figure 4.8. Curves drawn on the surface of the model, defining the pattern outlines* | *Figure 4.9: Mesh regions obtained from the boundaries created* | *back   front Figure 4.10. Resultant 2D patterns* |

The 2D patterns were analyzed in order to check whether the length of the seam-line was the same as that on the 3D model. As the length of the seam-line corresponded to that from the 3D model, the 2D patterns were further processed. The patterns for the upper front part of the legs were merged in order to have fewer seams while maintaining the same seam-line length as that in the original patterns. The shape of the pattern for the back of the knee was also improved in order to optimize the sewing procedure (Fig. 4.11). The new patterns were exported as a *.dxf* file and printed out in order to make the templates for the pants prototype.

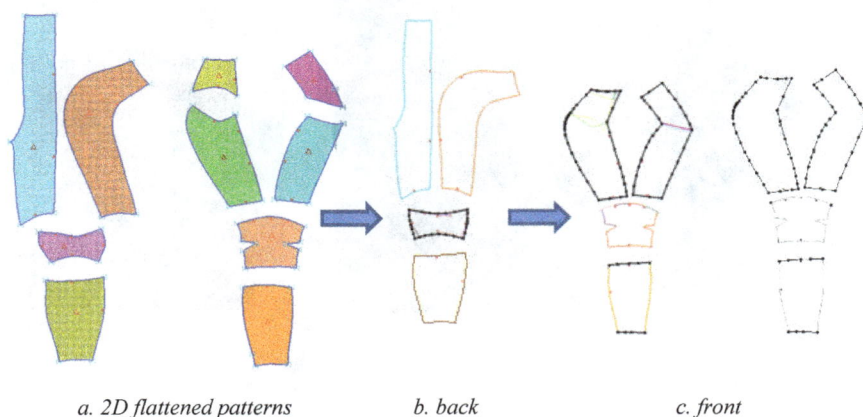

a. 2D flattened patterns   b. back   c. front

Figure 4.11. Modification of the flattened patterns (a) obtained for the back (b) and the front (c) of the pants

The pants prototype (Fig. 4.12) was made from an elastic fabric which is typically used for tight-fitting garments. It was seen, from the fit try-on, that the back was fully covered and that the pants followed the curve of the body in the sitting position, thanks to the patterns obtained from the 3D model. The front waistline was also in a good position, due to the decreased crotch length. The excess fabric in the groin area was eliminated in this way. Folds at the back of the knee were reduced, and the darts in the front area provided more room for the fabric to fit the knee-cap better.

The development of functional or comfortable garments for paraplegic persons, as stated in Chapter 2.3.2., has to avoid thick and hard seams; especially in areas which are exposed to high pressure-levels, such as the back and buttocks, that can lead to pressure-sores and skin lesions. It was thus necessary to design a tight-fitting pants model having fewer seam-lines for the buttocks and thigh patterns. Another posture, involving a minimal degree of bending, was chosen for this purpose in order to allow better flattening of the patterns. Starting from the first posture (with 90° and 110° bending angles) for which the pants model was created, up to the one needed to develop the new pants model, every increment of 10° in the bending angles was tested in order to obtain the best results during the flattening process. The angle of the upper body with respect to the thighs was changed from 90° to 130° and the angle at the knee was changed from 110° to 130° (Fig. 4.13).

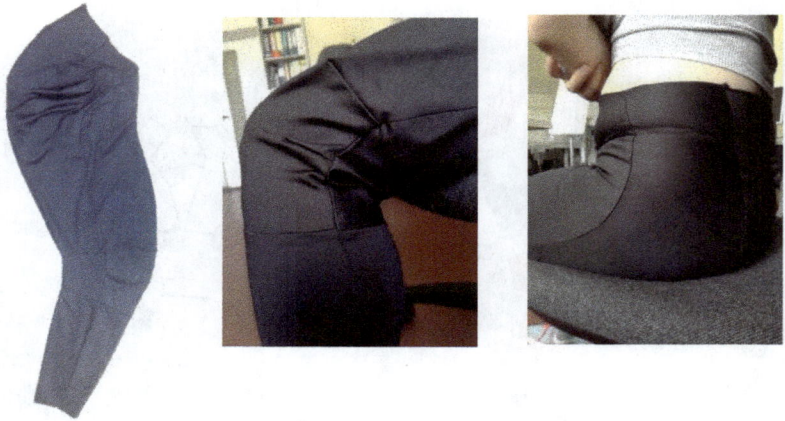

Figure 4.12: Trying-on of the tight-fitting pants prototype

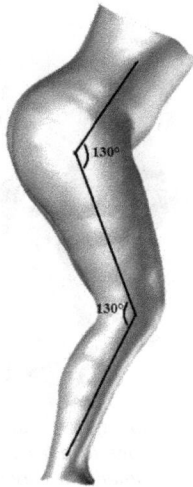

Figure 4.13: Trunk and knee bending angles for the second pants model

The new posture was chosen so as to give a 3D pants model having fewer seam-lines and thus make the flattening required to obtain good 2D patterns more attainable. The new tight-fitting pants model has only two patterns for the back; divided at the knee area, where the amount of fabric needs to be reduced. The front has three patterns; these being necessary in order to create a dividing area for the knee, where the fabric allowance has to be more generous. By applying *Draw curves › Create regions from curves › Flatten regions*, 2D patterns having a *Link length* of 9mm and a *Vertex angle* of 135° were obtained (Figs. 4.14 to 4.16). Because the bending angle was not as large as in the original position it was unnecessary to choose the smaller *Link length* of 6mm, given that the flattening process was already successful when using a value of 9mm.

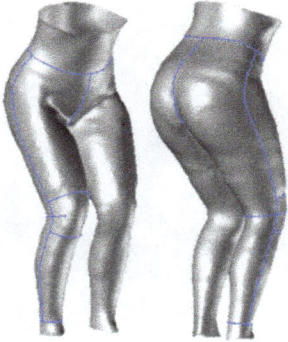

Figure 4.14. Curves drawn on the surface of the model, defining the pattern edges

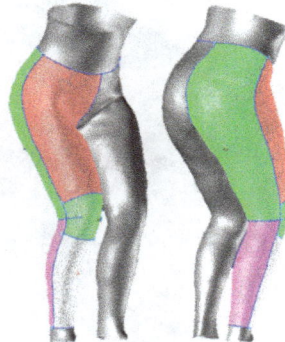

Figure 4.15. Mesh regions obtained from the boundaries created

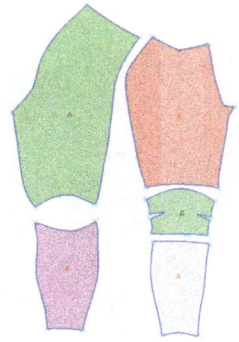

Figure 4.16. Resultant 2D patterns

Comparison with the previous patterns shows that, even though the bending angle was small, the pants model nevertheless furnished information concerning the modifications which are required for the areas of interest: i.e. buttocks, crotch and knees. Unnecessary seam-lines were eliminated, and information concerning the dimensional changes necessary in specific regions was obtained. The new patterns were tested further in order to establish rules for making the essential modifications to pants intended for wheelchair-users.

4.1.2   Fit simulation for the tight-fitting pants model

The patterns which were obtained by using the 3D-to-2D process were verified by virtual simulation on a 3D body-model. Garment simulation research is essential in many domains; such as, virtual clothing design, digital customization and e-commerce. Virtual fitting-simulation focuses on visualization of the garment on a virtual body-model in a 3D environment. The required software packages are available on the market (*Clo 3D* from *Clo Virtual Fashion* [176], *Modaris3D Fit* from *Lectra* [177], *OptiTex* from *EFI™* [178], *V-Stitcher 3D* from *Browzwear* [179]) and usually include the following: a 3D parametric mannequin[2] module, a fabric-properties module and a virtual pattern-sewing module [180]. The simulation process usually follows the same steps: it starts with the design of 2D

---

[2]term used in virtual fit simulation to replace that of 'virtual body model'.

patterns which are stitched and simulated on a 3D mannequin, while the draping effect is obtained by choosing various fabric characteristics from among those in the library [181]. These characteristics are fundamental to the draping effect of the garment, and the mechanical properties of the fabric will affect the performance of the clothing during wear-simulation with regard to deformation, constraints and allowances [182].

Although many studies have been made of virtual fit simulation for clothing the subject nevertheless remains challenging when it comes to obtaining a realistic fit simulation. If the mechanical properties of fabrics available in the software library cannot offer an accurate and realistic representation, of deformation and of interaction of the fabric with the virtual space, the virtual representation can be affected [131]. Another problem could be that the library of virtual mannequins usually contains standard body-types which are posed in a standardized upright position, thus making simulation more difficult when it is required to use a different body position. The fit-simulation process and efficiency may therefore vary according to the type and style of the garment, the fabric characteristics and the shape of the mannequin. The simulation software can usually import specific fabric data and personalized 3D body models, which is helpful in research on specific clothing simulations for different fabrics and body positions.

Virtual fit simulation was performed using the *ModarisV8R1* program in the case of the 2D patterns obtained using the 3D-to-2D methodology. *Modaris* is software that is dedicated to model-making, pattern-making, industrialization and grading. *Modaris 3D Fit* permits the virtual prototyping of 2D-designed patterns. It comprises a virtual assembly module, a mannequin module, 3D fabric selection from a library and 3D simulation; with the possibility of analyzing mesh, cross-line and grain-line deformations and fabric easing. The *.dxf* files of the 2D patterns were then imported, arranged and mirrored in order to perform simulations of the entire pants model (Fig. 4.17).

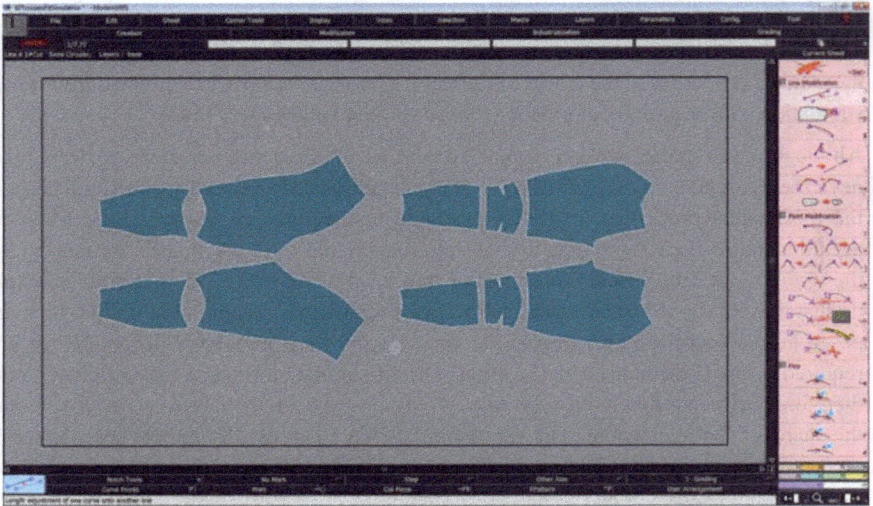

Figure 4.17. Modaris work area for 2D patterns [source Lectra]

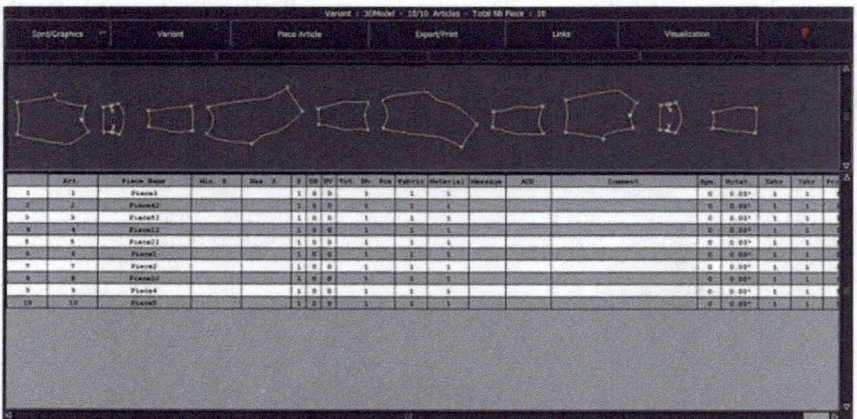

Figure 4.18. Model variant [source Lectra]

The model variant for the 3D simulation is designed using (*F8*): *Variant › Create Variant (Insert Variant name) › Create Piece Article* (Fig. 4.18). The stitch lines are declared in *F~: Desk of Stitches › Add Stitch* (Fig. 4.19).

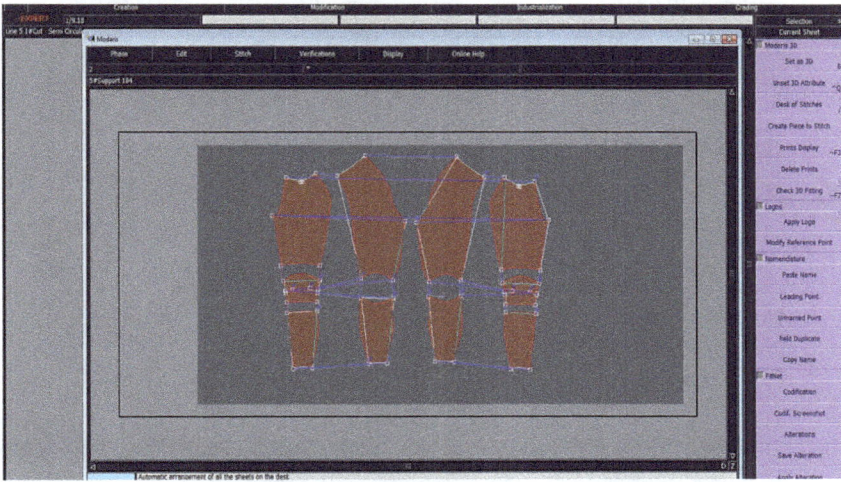

Figure 4.1. Desk of stitches, [source Lectra]

Using *Check 3D Fitting* from the *~F1* function, the patterns are transferred to the simulation program and the virtual mannequin, developed in *3dsMax*, is imported. The seam-lines have to be switched for the working area. Standard seam-lines will appear in the same color: command bar -*Garment - All seams* (Fig. 4.20). The next step is to choose the fabric characteristics: command bar - *Material - Fabric*. Denim made from 100% cotton, with a weight of 385g/m^2, was added by using the existing fabric library. The mechanical characteristics are presented in Table 4.1 in accord with the FAST (Fabric Assurance by Simple Testing) system.

Table 4.1. Fabric characteristics according to the FAST system

Cotton Denim [100% cotton]			
Density 385 [g/m²]			
Fabric thickness [0.08cm]			
Mechanical characteristics		warp	weft
Bending rigidity– B [Nm]		97.4	38.3
Fabric elongation [g/cm²]	E5 %	0.0093	0.078
	E20 %	0.15	0.28
	E100 %	1.03	1.68

Where: g/m² = grams per square meter
      Nm = Newton-meter
      g/cm² = grams per square centimeter
      E = elongation

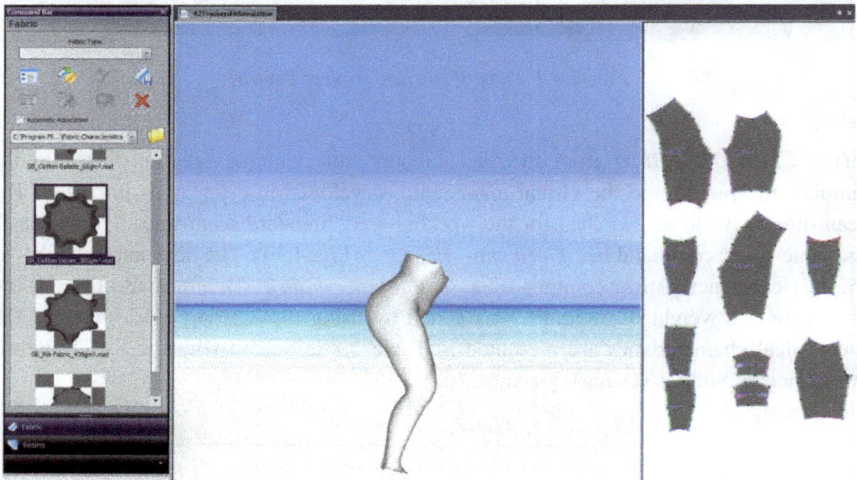

Figure 4.20. Modaris 3D work area with the mannequin and defined seam-lines on the 2D patterns [source Lectra]

The Australian CSIRO Division of Wool Technology developed the FAST system, which comprises a set of instruments and methods for measuring the mechanical and dimensional characteristics of fabrics. Fabric performance and the appearance of the garment can be tested with the help of FAST while it is worn. The system involves tests using three measuring instruments [184,185, 186]:

➤ Fast 1: is a compression meter that measures the thickness of the fabric under two fixed loads. The fabric is first measured using a load of $2g/cm^2$ and is then again tested using a load of $100g/cm^2$;

➤ Fast 2: is a bending meter that is used to measure the stiffness, or conversely the flexibility, of a fabric; the instrument works on the cantilever principle which involves pushing fabric over a vertical edge until it has bent to a specified angle (41.5°). The bending length is then converted into a bending rigidity which is directly related to fabric stiffness;

➤ Fast 3: is an extension meter that measures the percentage stretching of a fabric under three low (5, 20, $100g/cm^2$) fixed loadings. Fabrics are measured under all three loads in the warp and weft directions and, under the lowest load only, in the 45° bias direction. The latter extension is used to calculate the shear rigidity, which is directly related to fabric looseness.

Interactive pre-positioning of the patterns was necessary (Fig. 4.21) for pants simulation on the mannequin. In the *Assembly* command bar, the *Simulate* process was applied and the patterns were sewn together on the mannequin (Fig. 4.22). It can be seen that the patterns for the back do not fit well. The mesh surface in the curved area can suffer some modifications during the 3D-2D flattening process because of inaccuracies in the flattening of such curved pattern pieces. The differences between the 3D surface and 2D surface can vary according to the degree of curvature. Table 4.2 lists the differences in cm^2 between the 3D and 2D patterns. Values marked with a "-" exhibit a more significant surface area in the 2D patterns.

Inaccuracies in the flattening process can be noticed on the simulated pants model with regard to the deficit of fabric from the back. The location of the pattern for the knees also fails to satisfy (Fig. 4.23). Although they provide valuable information concerning the modifications the patterns are nevertheless unsuccessful for this type of posture; the 3D-to-2D technique cannot be used to obtain tight-fitting pants for the bent posture directly, because of the flattening-process error. The 2D patterns were further modified by increasing the width in order to rectify the problem identified in the 3D representation. It was necessary to add easing values in the waist and hip perimeters and in the knee and pants hems (Fig. 4.24).

Table 4.2. Differences between the pattern surface areas, 3D versus 2D

	3D patterns (cm²)	2D patterns (cm²)	Difference 3D-2D (cm²)
Upper side	1615.6	1521.7	93.9
Lower back	570.3	546.1	24.2
Upper front	1198	1220.9	-22.9
Knee	257.3	246.1	11.2
Lower front	434.9	449.2	-14.3

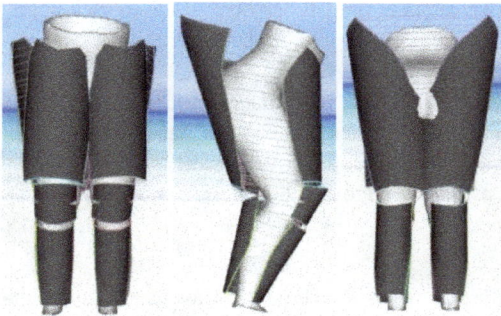

Figure 4.21. Interactive pre-positioning of the patterns

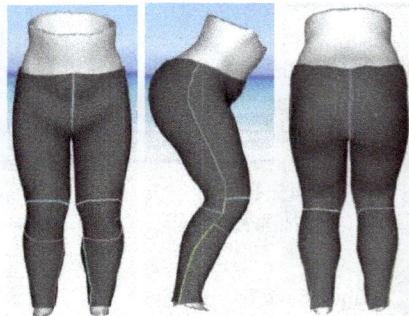

Figure 4.22. Simulated pants

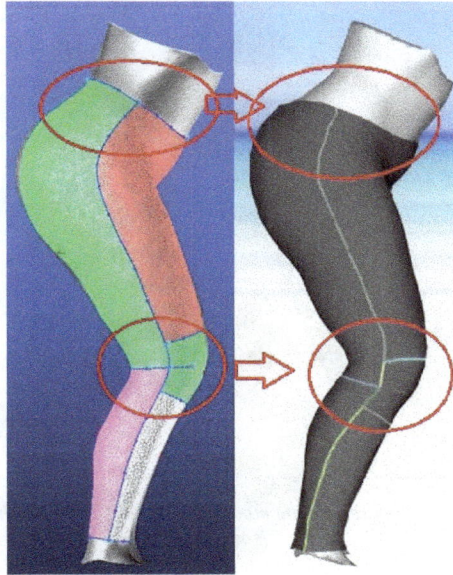

Figure 4.23. Differences in fit between the 3D pants model (left) and the simulated flattened patterns (right)

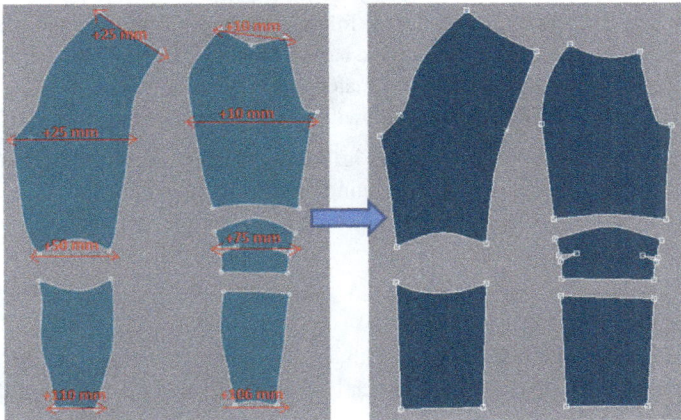

Figure 4.24. Ease modification of the patterns obtained using the 3D-2D flattening procedure [source Lectra]

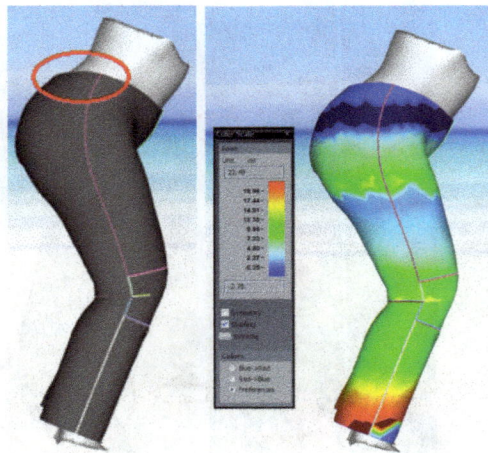

Figure 4.25. Virtual simulation of the modified 3D pants model with fabric ease distribution [source Lectra]

The fit simulation followed the same procedure as that described earlier. The patterns were stitched together in *Modaris* and then exported to *Modaris 3D.* The seam-lines were defined, the same fabric was chosen and, following pre-positioning of the patterns, the *Simulate* process on the *Assembly* command bar was applied. Although a better fit was thus obtained, the knee-line was adequately situated and the ease distribution led to excellent fitting with regard to comfort, the problem with the back remained unresolved (Fig. 4.25).

The color-coding of the ease-simulation (Fig. 4.25) indicates negative values in the dark-blue colored area, and positive values ranging from blue to green and red. The negative values show a lack of easing of the product on the body, while the positive ones indicate areas where the product has ease.

The 2D patterns for the bent posture require some modifications in order to ensure better fitting to the body:

- a more significant back crotch length so as to cover the back correctly, making it necessary to measure the back crotch length for this position for use in the pattern construction process;
- the length of the front crotch had to be shortened in order to ensure proper fitting at the waistline;

- the amount of fabric in the back-of-knee area had to be reduced in order to avoid excessive folding that could irritate the skin;
- the allowance of fabric for the front of the knee needed to be greater in order to offer better comfort when in the sitting posture.

Although it provides valuable information with regard to pattern modification, the 3D-to-2D technique again cannot be used without modification in order to obtain tight-fitting pants for a bent position directly because of the flattening-process error. The fitting of the 3D pants model was therefore analyzed further in order to see whether it was possible to obtain customized loose-fitting pants by applying the modifications resulting from use of the 3D-to-2D technique.

4.2   2D-3D virtual prototyping

The 2D-to-3D virtual prototyping involves, as mentioned before, 3D fit simulation of 2D-designed patterns. The patterns are first conventionally designed in this manner using CAD programs and are then imported into 3D simulation programs. A 2D basic pants model was designed according to the required pattern-construction specifications. The patterns were further modified as guided by changes suggested by the flattened patterns of the 3D pants model. Fittings of both the basic pants and the modified ones were analyzed for the bent position.

4.2.1   The construction of basic pants patterns and fit simulation

The basic patterns comprise the planar pieces that form any type of garment, and depend in form upon specific aspects of body-shape and clothing-style. The basic pattern is developed by performing calculations which are based upon body-size measurements, size-charts derived from anthropometric studies, grading increments deduced from both body measurements and size charts and various types of allowances [186]. The 2D construction of the basic patterns for women's pants was carried out while following measurements, for the construction of outerwear for ladies, which were based upon the size-charts of the DOB - Verband, using the _Modaris_ program from _Lectra_. The necessary line segments, curves or circles were calculated by using specific mathematical formulations taken from the construction-specifications chart which defined body dimensions such as the in-seam or side-seam length, crotch length, waist or hip line etc. Upon completing the entire network of necessary lines for the patterns, the front and rear pieces were extracted (Fig. 4.26), and used for the virtual fit simulation.

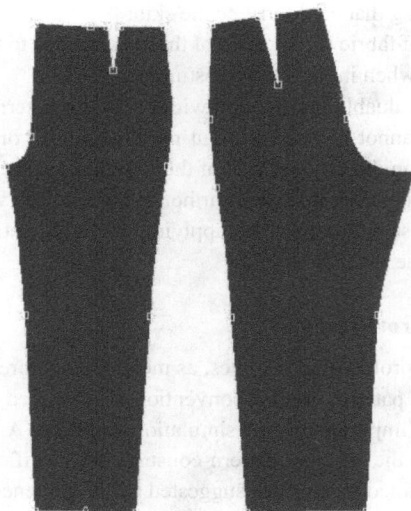

Figure 4.26. Front and basic back patterns for women's pants [source Lectra]

The virtual fit simulation was performed on the standing position developed in *3dsMax* (Fig. 4.27) in order to check the 2D basic patterns which were obtained. The fitting results showed that the resultant pants model was suitable for the standing position. The basic pants fit simulation was next performed for the bent posture, with an angle at the knees and trunk area of 130° (Fig. 4.13). From the virtual fit results (Fig. 4.28) it can be seen that a basic pants model, for a bent body position, requires some modifications; otherwise the rear of the body remains uncovered and the front waistline of the pants rises above the natural waistline. Some extra folds also appear at the back of the knees; all of these aspects could cause discomfort to a wheelchair-user.

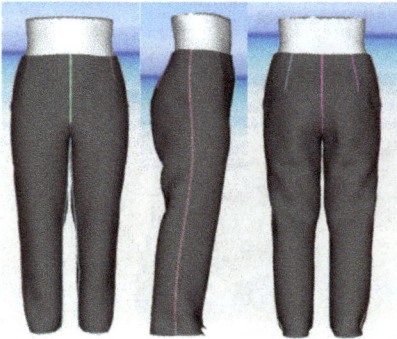

Figure 4.27. Fit simulation of basic pants in the standing position – front, side and rear views

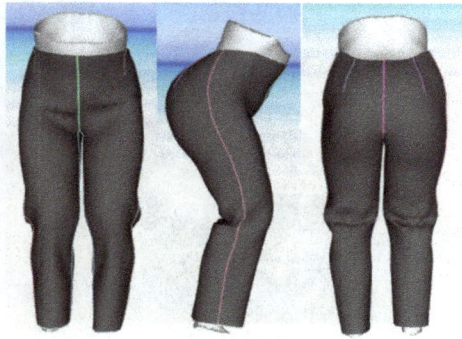

Figure 4.28. Fit simulation of basic pants in the bent position with 130° bending of the knees and trunk – front, side and rear views

4.2.2   Modification of the basic pants patterns

It was concluded on the basis of the fit simulation results that the basic patterns had to be modified at the rear by extending the crotch. The length of the front crotch must be shortened in order to ensure good comfort for the wearer. The problem with the folds at the back of the knee also has to be solved. Dimensional analysis of the 3D and classic procedures was carried out in order to modify the basic patterns on the basis of these conclusions. The length of the rear crotch for the 3D pants model and the basic model is approximately the same, with just a slight difference of 0.25cm between them (Fig. 4.29). The fit simulation showed that both led to a problem in covering the rear in the bent position. The distance from the pants waistline to the natural one was measured in order to determine by how much the crotch length had to be extended. (Fig. 4.30). The length of the rear crotch of the basic pants had to be increased by 10cm in order to fit the body well in this particular case. It was also necessary to modify the shape of the pants in order to solve the problem of the folds at the back of the knee. The patterns which were obtained by using the 3D model were measured (Fig. 4.31). A reduction of 6cm had to be made at the back of the knee in order to fit well the bent position of the body. The basic pattern for the rear of the pants was divided into two at the knee line, and a reduction of 3cm was made on each side.

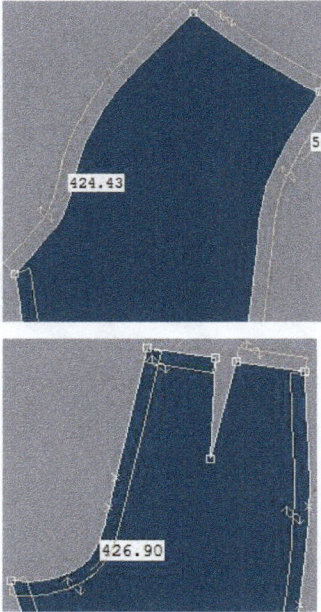

Figure 4.29. Length in mm of the rear crotch for 3D (top) and basic pants model (bottom) [source Lectra]

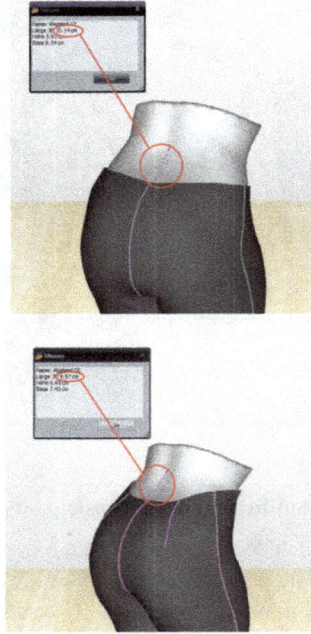

Figure 4.30: Distance from the rear waistline of the pants to the natural waistline of the mannequin body for the 3D model (top) and the basic model (bottom)

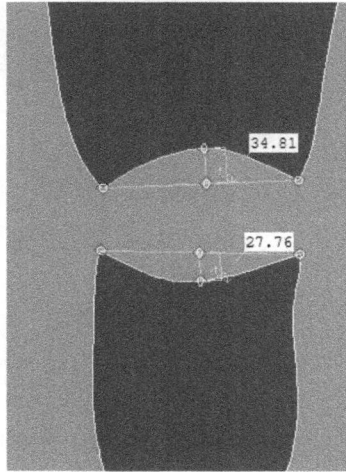

Figure 4.31: Dimensions (mm) of the back of the knee for the 3D pants model [source Lectra]

The fit simulation of the 3D model for the front gave good results, and the waistline was correctly positioned to offer comfort in the bent position. The basic pants simulation showed that the pattern had to be modified so as to decrease the front length of the crotch and give a better fit at the waistline. It was necessary to determine the length of the crotch from the 3D model, and that from the basic model, for the purpose of front-crotch modification (Fig. 4.32). The depth of the front dart was used to model the front pattern. Two darts gave extra ease in the front knee area (Fig. 4.33).

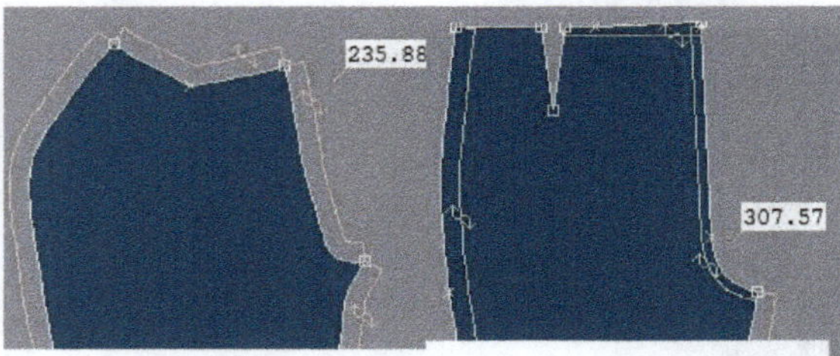

Figure 4.32. Length (mm) of the front crotch for the 3D model (top) and basic model (bottom), [source Lectra]

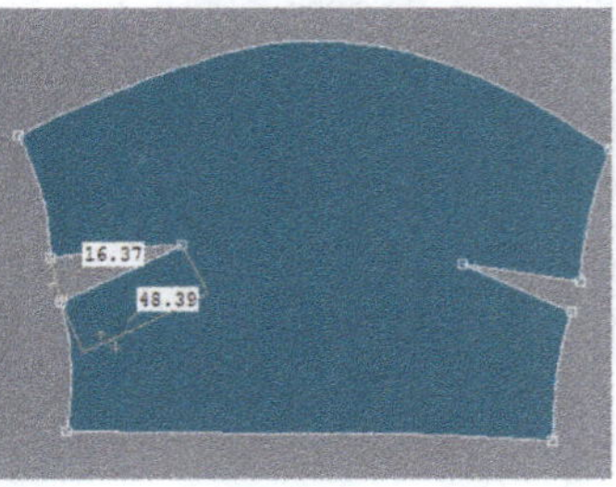

Figure 4.33. Knee dart measurement on the 3D model, [source Lectra]

All of the steps followed in the pattern modification are presented in Figs. 4.34 - 4.42.

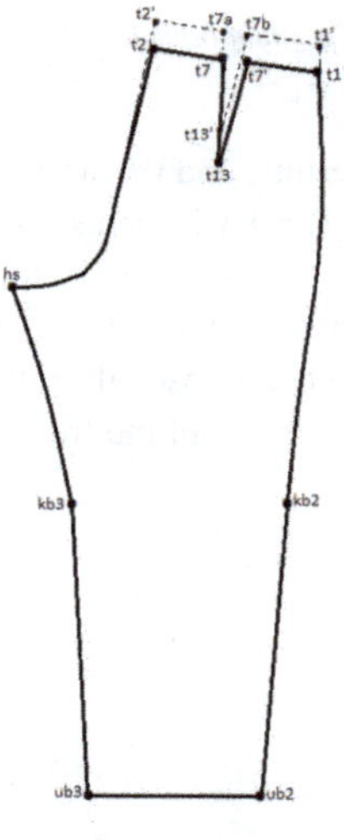

Figure 4.34. Rear modification 1

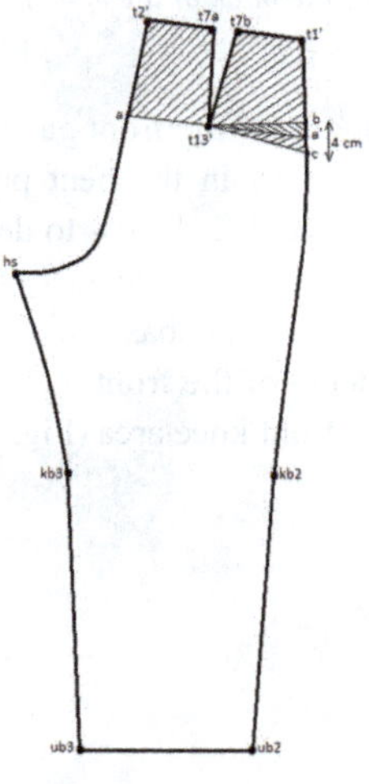

Figure 4.35. Rear modification 2

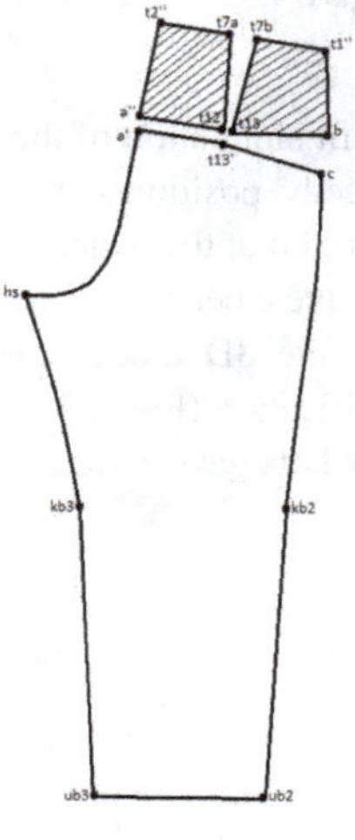

Figure 4.36. Rear modification 3

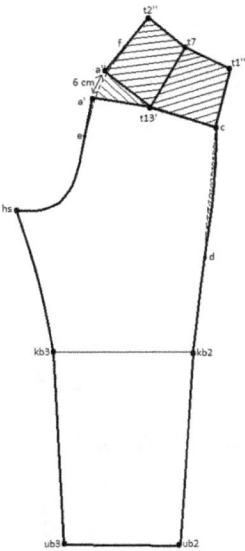

Figure 4.37. Rear modification 4

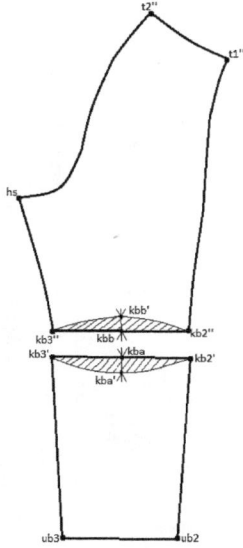

Figure 4.38. Rear modification 5

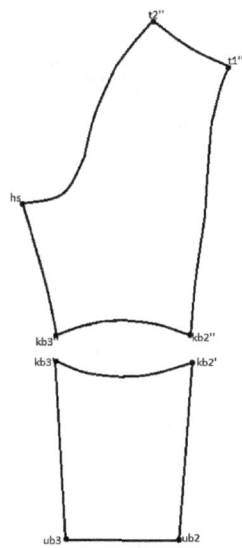

Figure 4.39. Rear modification 6

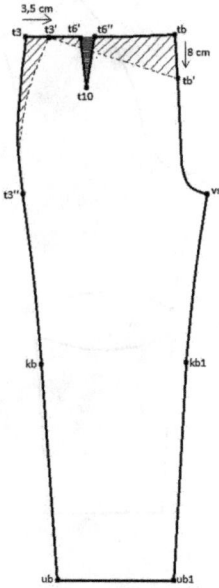

Figure 4.40. Front
modification 1

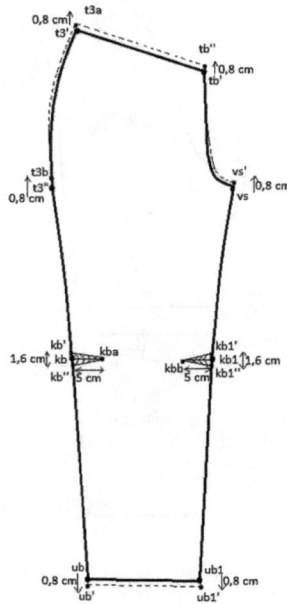

Figure 4.41. Front modification 2

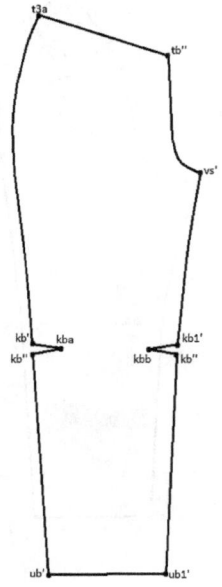

Figure 4.42. Front
modification 3

The procedure which was used to transform the patterns for the basic pants followed logical
and calculated steps. The new pants model combines the features of loose-fitting pants with
the shape and dimensional characteristics of the tight-fitting pants model (Figs. 4.43 and
4.44).

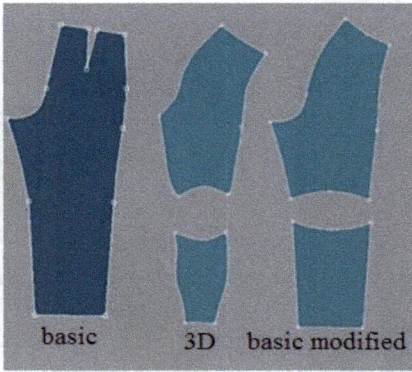

Figure 4.43. Rear views of the three pattern models [source Lectra]

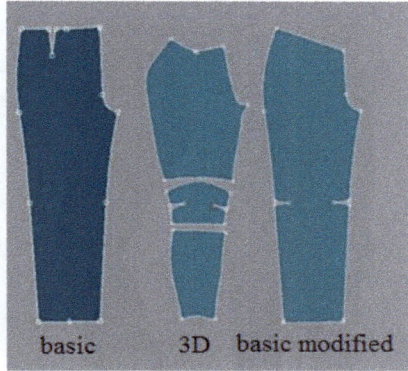

Figure 4.44. Front views of the three pattern models [Source Lectra]

4.2.3   Fit simulation of the modified basic pants

The modified pants model was verified virtually. The fit simulation was performed by following the same steps as those described in Chapter 4.1.2. Analyzing the obtained fit result (Fig. 4.45) shows that the new pants model gave a good fit for the waist area. The back was covered, and the front waistline was in an excellent position to offer comfort in the bent body position. The small fold which appears at the back of the waistline can be adjusted by using an elastic section that was not included in the simulation process. Modification of the pattern at the back of the knee reduced the number of fabric folds. According to the fabric ease distribution (Fig. 4.46) the pants offer an excellent degree of fit. The waistline and hip area, with a proper easing allowance (dark blue area) made, impart a balanced fit of the pants to the body. The crotch area increased the ease allowance and offered increased comfort for the body (blue to green area).

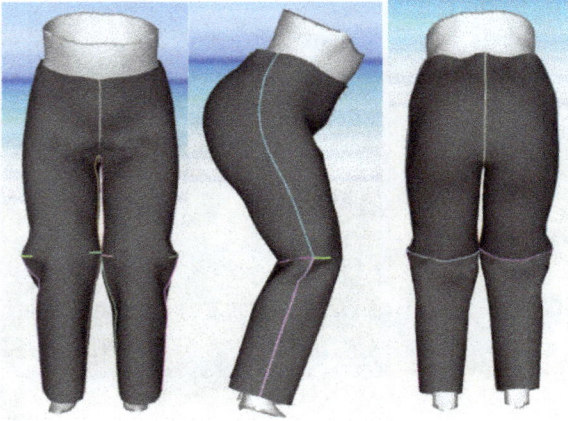

Figure 4.45. Fit simulation of the modified basic pants

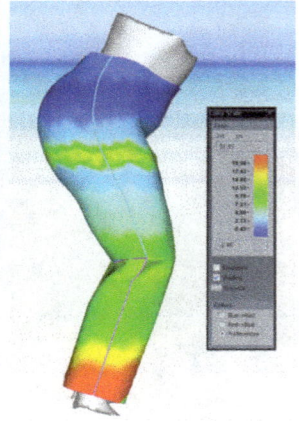

Figure 4.46. Fabric ease distribution for the modified basic pants

4.3   Designing a pair of pants for a wheelchair-user

Based upon the studies presented in Chapter 4.2 the same principles and stages were applied to the design of a pair of pants for a wheelchair-user. Analyzing the measurements taken in the virtual environment (Fig. 4.47), and directly on the body, of the wheelchair-user presented in Chapter 3.1.4 it can be seen that the pelvic area is the most difficult one for which to obtain precise dimensions (Table 4.3). The hip circumference and the crotch length exhibit the most significant differences between the virtual and the classic measurements.

Table 4.5: Measurements (cm) of the body of a wheelchair-user

Measurement	I. Scanning method	II. Classical method	Difference I. −
Waist circumference	113.6	112	1.6
Hip circumference	111.4	106	5.4
Thigh circumference	47.1	49	-1.9
Knee circumference	43.2	42	1.2
Calf circumference	29.3	29	0.3
Ankle circumference	25.7	26	-0.3
Crotch length	76.3	80	-3.7

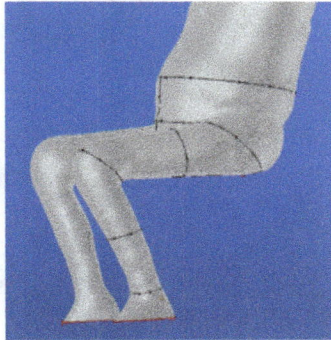

Figure 4.47. 3D virtual measurement locations for the wheelchair-user

The hip circumference of the natural body is problematic with regard to adjustment of the tape-measure so that it is in the same position on the buttocks as it is in the virtual environment. The paraplegic person can support himself with his arms so as to facilitate hand measurements using a tape, but he cannot maintain that posture for any great length of time.

The crotch length is also problematic because of the contact area; the tape cannot be placed in the same position, due to intimacy issues. The differences between the other measurements are quite slight. The smaller value found for classic measurements of the waist area may be due to the inability of the person to breathe in a relaxed manner. Differences in the thigh and knee areas can occur because of the mesh modeling that is performed in order to repair areas which are missing from the scan data.

The next step was to take a basic pants model for men and apply the pattern modifications which had previously been made to pants for women. The modifications were made by following the same steps as those described in Chapter 4.2.2, and with the same dimensional changes. By comparing the hip area measurements, obtained for the wheelchair-user, with the size chart of dimensional specifications for clothing designed for men a basic pants model (Fig. 4.48) was created and a further modified prototype could be tried-on by the wheelchair-user. The knee line has to be lowered in order to ensure a better fit to the body.

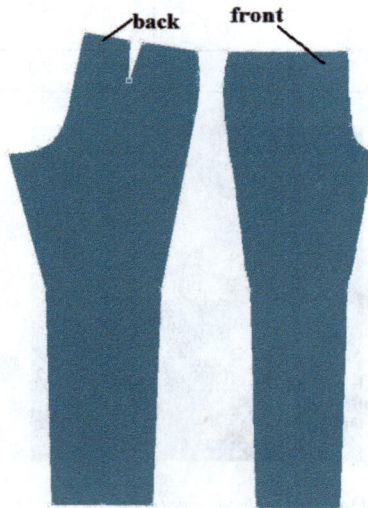

Figure 4.48: Back and front patterns for men's pants model [source Lectra]

All the steps followed for pattern modifications are presented in Figs. 4.49 - 4.57.

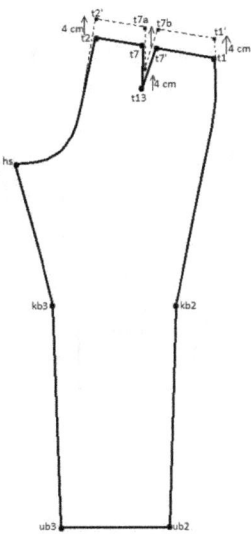

Figure 4.49. Rear modification 1

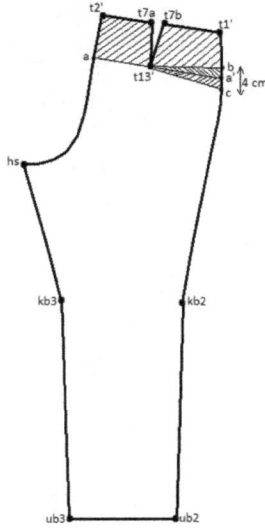

Figure 4.50. Rear modification 2

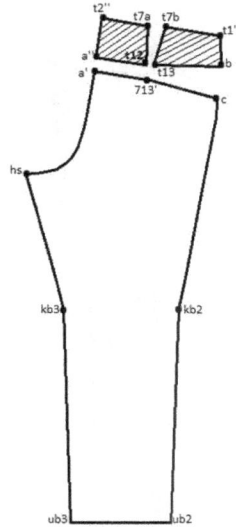

Figure 4.51. Rear modification 3

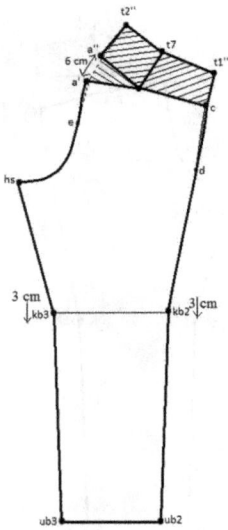

Figure 4.52. Rear
modification 4

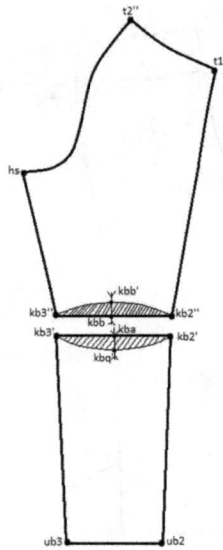

Figure 4.53. Rear
modification 5

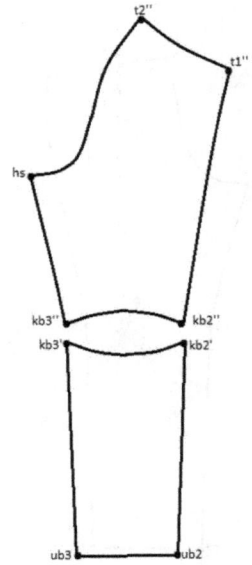

Figure 4.54. Rear
modification 6

Digital Methods in Developing Textile Products for People with Locomotor Disabilities
Materials Research Foundations **110** (2021)

Materials Research Forum LLC
https://doi.org/10.21741/9781644901557

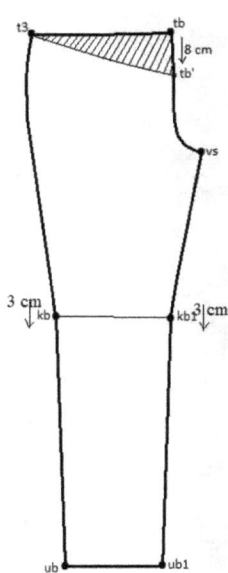

Figure 4.55. Front modification 1

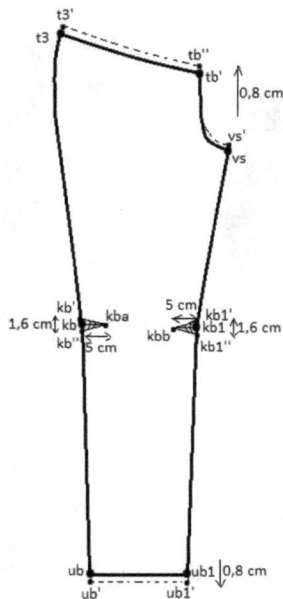

Figure 4.56. Front modification 2

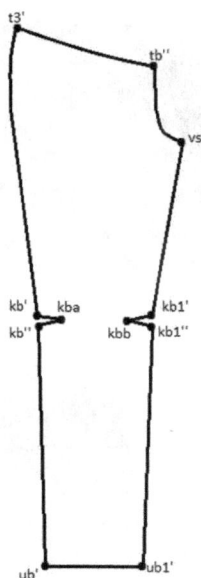

Figure 4.57. Front modification 3

The resultant patterns were further printed, prepared for sewing and the final product then tried on by the wheelchair-user (Fig. 4.58, Fig. 4.59). A zipper having the same length as the front crotch was chosen for the prototype so that dressing was facilitated. Elastic tape was chosen for the waistband so as to secure the pants on the body and prevent slipping when the subject had to move from the chair. The pants prototype fitted well with regard to comfort. The subject said that he prefers a lowered front waist, with a more prominent belly because of his personal body conformation, but that this was not a significant inconvenience for him. This will not be true for all wheelchair-users because of their particular preferences and body morphologies. The back was fully covered and the knee-line fitted well, with reduced folding at the back of the knee. The fact that the pants were loose-fitting was considered an exemplary aspect, as it could hide the deformity of an atrophied leg.

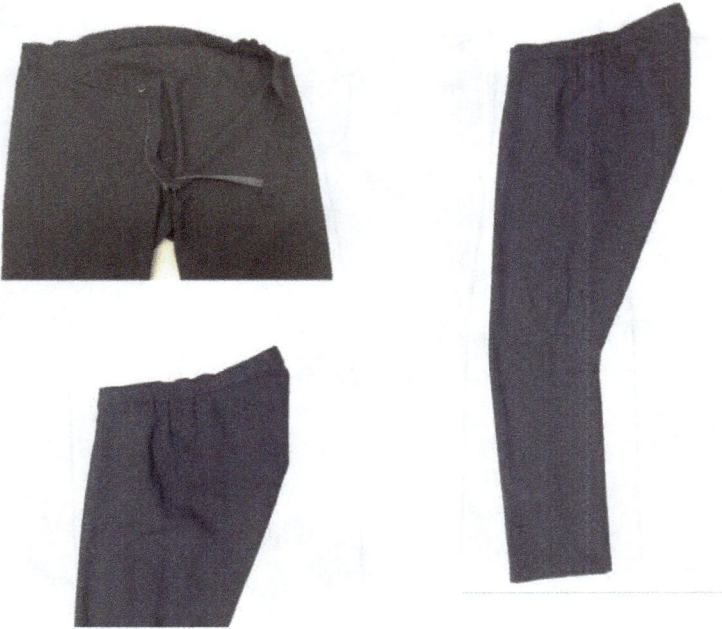

Figure 4.58. Modified pants prototype

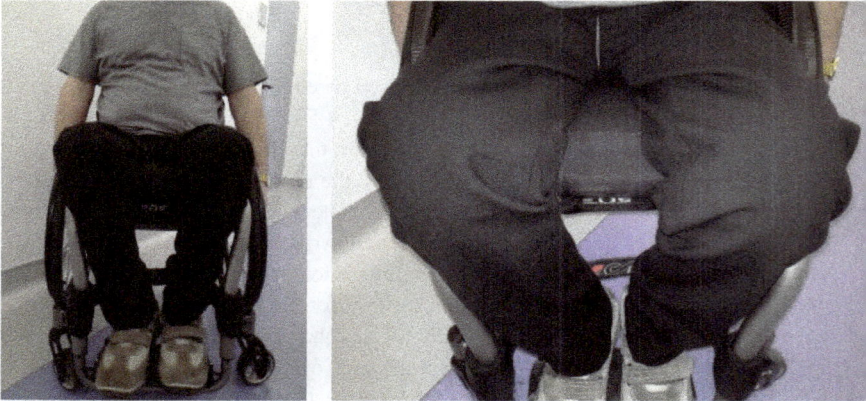

Figure 4.59. Wheelchair-user wearing the modified pants model

When it comes to the fabric used the subject prefers, and chooses for day-to-day living, jeans pants made from 100% cotton as this lets the skin breath properly. The fabric also has a lower elasticity and this helped him to move his legs from one place to another. In response to this request a pair of jeans was designed, following the same modification rules, using a 100% cotton fabric (Fig. 4.60).

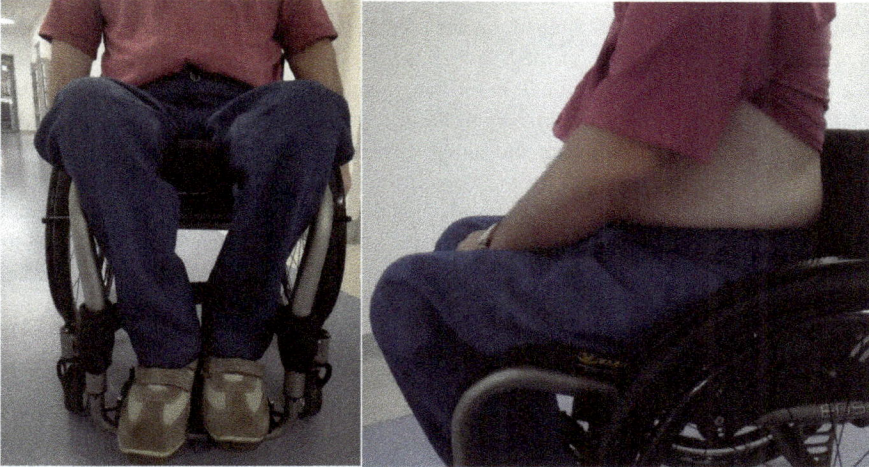

Figure 4.60. Wheelchair-user wearing the final jeans model

Conclusions

The primary purpose of analyzing the 3D design method, presented in this study, was to implement 3D CAD programs for the development of functional pants for wheelchair-users. The 3D-to-2D method proved to be successful in obtaining the required information concerning the dimensional changes which a pair of pants for the sitting position should incorporate; designing a 3D tight-fitting pants model permitted analysis of the necessary modifications to be made to the patterns for a basic loose-fitting pants model. Following modification of the patterns for the basic pants a virtual fit simulation demonstrated that the applied changes improved the fit in the bent position. Starting with the positive results obtained for the modified virtual fitting of basic pants for women, it was natural to create a prototype that a wheelchair-user could accept. The modifications which had been applied to the pants model for women proved to be relevant to the pants for men; with just some small personal recommendations by the wheelchair-user. It can be concluded with regard to fitting that the pattern-modification method is valid for designing functional pants for people with paraplegia. Further studies might be conducted in order to establish a standard pattern-design method for creating functional pants for this group of people, with their differing body-types and sizes.

The pants patterns for wheelchair-users were designed with the aid of the commands of 3D CAD programs. Detailed procedures were analyzed and described with regard to finding the best possible solutions for obtaining anthropometric information on wheelchair-users and creating a standard method for the design of pants models for this group of people. The resultant methods described in this study open up new research directions that can be further elaborated. For example:

➢ for the scanning procedure a specific protocol could be established, according to the disability of the subject, that would ensure accuracy in obtaining information for advanced anthropometric studies;

➢ taking into account the importance of the comfort of disabled people regarding their clothing, fabric properties need to be researched further in order to improve and develop optimum materials for this type of product, and

➢ a technical design method could be developed for the pants model and correlated with fabric properties such as elasticity, density and stiffness in order to impart optimum comfort to the product during its wearing.

References

[1]   World Health Organization, World Report on Disability - Summary, World Rep.
        Disabil., no. WHO/NMH/VIP/11.01, pp. 1–23, 2011.

[2]   Ministry of Labor and Social Justice and National Authority for Disabled Persons,
        Statistic Data, 2018.

[3]   K. E. Carroll and D. H. Kincade, Inclusive Design in Apparel Product
        Development for Working Women With Physical Disabilities, Fam. Consum. Sci.
        Res. J., vol. 35, no. 4, pp. 289–315, 2007.
        https://doi.org/10.1177/1077727X07299675

[4]   A. Kabel, J. Dimka, and K. McBee-Black, Clothing-related barriers experienced by
        people with mobility disabilities and impairments, Appl. Ergon., vol. 59, pp. 165–
        169, 2017. https://doi.org/10.1016/j.apergo.2016.08.036

[5]   S. Kirshblum and M. Brooks, Neurological Examination and Classifi cation in
        Spinal Cord Injury, in Practical Psychology in Medical Rehabilitation, S. I.
        Publishing, Ed. Switzerland 2017, 2017, pp. 33–40. https://doi.org/10.1007/978-3-
        319-34034-0_5

[6]   J. V. Durá-Gil et al., New technologies for customizing products for people with
        special necessities: project FASHION-ABLE, Int. J. Comput. Integr. Manuf., vol.
        30, no. 7, pp. 724–737, 2017. https://doi.org/10.1080/0951192X.2016.1145803

[7]   World Health Organization, International Classification of Impairments,
        Disabilities, and Handicaps, Geneva, 1980.

[8]   World Health Organization, Disabilities, Information on:
        http://www.who.int/topics/disabilities/en/.

[9]   World Health Organization, International Classification of Functioning , Disability
        and Health, 2001.

[10]   World Health Organization, Raport mondial privind dizabilitatea/ World Disability
        Report, Bucuresti, 2012.

[11]   V. M. Alford, L. J. Remedios, G. R. Webb, and S. Ewen, The use of the
        international classification of functioning, disability and health (ICF) in indigenous
        healthcare: A systematic literature review, Int. J. Equity Health, vol. 12, no. 1, p. 1,
        2013. https://doi.org/10.1186/1475-9276-12-32

[12]   N. Kostanjsek, Use of the International Classification of Functioning, Disability

and Health (ICF) as a conceptual framework and common language for disability statistics and health information systems, BMC Public Health, vol. 11, no. SUPPL. 4, p. S3, 2011. https://doi.org/10.1186/1471-2458-11-S4-S3

[13] M. Comanescu, C. Predescu, and M. Stanciu, Abilitățile Contează – Implementarea Clasificării Internaționale a Funcționării, Dizabilității și Sănătății în serviciile de incluziune socială pentru persoanele utilizatoare de scaune rulante din România, Swiss-Romanian cooperation programme, Sovis Print, Information on: www.motivation.ro www.paraplegie.ch

[14] R. L. Metts, Disability Issues , Trends and Recommendations for the World Bank, pp. 1–91, 2000.

[15] H. Council and C. Policy, Needs analysis: adults with physical disabilities, no. October, 2007.

[16] A. Curteza, V. Cretu, L. Macovei, and M. Poboroniuc, Designing functional clothes for persons with locomotor disabilities, Autex Res. J., vol. 14, no. 4, pp. 281–289, 2014. https://doi.org/10.2478/aut-2014-0028

[17] C. S. Hawke, Accommodating students with disabilities, New Dir. Community Coll., vol. 2004, no. 125, p. 17, 2004. https://doi.org/10.1002/cc.141

[18] K. E. Carroll, (2002). Innovations and improvisations: A study in specialized product development focused on business clothing for women with physical disabilities (Order No. 3147777). Available from ProQuest Dissertations & Theses Global. (305519881). Retrieved from https://proxying.lib.ncsu.edu/index.php/login?url=https://www-proquest-com.prox.lib.ncsu.edu/dissertations-theses/innovations-improvisations-study-specialized/docview/305519881/se-2?accountid=12725. Original definitions obtained from Encyclopedia of Disability and Rehabilitation by A. E. Dell Orto, and R. P. Marinelli, (1995).

[19] T. M. Dixon and M. A. Budd, Spinal Cord Injury, in Practical Psychology in Medical Rehabilitation, Switzerland: Springer International Publishing, 2017, pp. 127–136. https://doi.org/10.1007/978-3-319-34034-0_15

[20] V. Sabapathy, G. Tharion, and S. Kumar, Cell Therapy Augments Functional Recovery Subsequent to Spinal Cord Injury under Experimental Conditions, Stem Cells Int., vol. 2015, p. 12, 2015. https://doi.org/10.1155/2015/132172

[21] R. N. Auer, M. Oehmichen, and H. G. König, Injuries of Spine and Spinal Cord, in Forensic Neuropathology and Associated Neurology, Germany, 2006, p. 219.

https://doi.org/10.1007/3-540-28995-X

[22] W. Y. Yu and D. W. He, Current trends in spinal cord injury repair, Eur Rev Med Pharmacol Sci, vol. 19, no. 18, pp. 3340–3344, 2015.

[23] I. Bromley, Tetraplegia and Paraplegia, a guide for physiotherapists. Ltd, Elsevier, 2006.

[24] M. Pazzaglia, G. Galli, G. Scivoletto, and M. Molinari, A Functionally Relevant Tool for the Body following Spinal Cord Injury, PLoS One, vol. 8, no. 3, pp. 1–8, 2013. https://doi.org/10.1371/journal.pone.0058312

[25] Anatomy of the Spine. Retrieved from: https://sites.google.com/a/wisc.edu/neuroradiology/anatomy/spine/slide-1.

[26] M. Brooks and S. Kirshblum, Spinal Cord Injury, in Essential physical medicine and rehabilitation, New Jersey: Humana Press Inc, 2006, pp. 59–100. https://doi.org/10.1007/978-1-59745-100-0_3

[27] World Health Organization, International Perspectives on Spinal Cord Injury. Malta, World Health Organization, The International Spinal Cord Society, 2013.

[28] D. Fasouli, Perception of Paraplegia (Reaction of a Typical Person towards a Paraplegic Person, the Role of Knowledge and Experience), no. March 2013, 2014.

[29] V. R. Preedy and R. R. Watson, Cauda Equina Syndrome, Handbook of Disease Burdens and Quality of Life Measures. Retrieved from: https://link.springer.com/content/pdf/10.1007%2F978-0-387-78665-0_5263.pdf.

[30] W. P. Howlett, Clinical Skills, in Neurology in Africa, 2012.

[31] T. Miron, Paraplegia. Information on: https://www.prostemcell.org/leziuni-ale-coloanei-vertebrale/paraplegia.html.

[32] A. Marx, J. D. Glass, and R. W. Sutter, Differential diagnosis of acute flaccid paralysis and its role in poliomyelitis surveillance., Epidemiol. Rev., vol. 22, no. 2, pp. 298–316, 2000. https://doi.org/10.1093/oxfordjournals.epirev.a018041

[33] Difference Between Flaccid and Spastic Paralysis. Retrieved from: http://www.differencebetween.com/difference-between-flaccid-and-vs-spastic-paralysis/.

[34] G. Vyshka, A. Kuqo, S. Grabova, E. Ranxha, L. Buda, and J. Kruja, Acute flaccid paraplegia : neurological approach, diagnostic workup , and therapeutic options, J. Acute Dis., vol. 4, no. 1, pp. 1–6, 2015. https://doi.org/10.1016/S2221-

6189(14)60073-1

[35] R. Schüle and S. Ludger, Spastic Paraplegia, Hereditary, in Encyclopedia of Molecular Mechanisms of Disease, 2009, pp. 1952–1956.

[36] D. Guptaa, Functional clothing- definition and classification, Indian J. Fibre Text. Res., vol. 36, no. 4, pp. 312–326, 2011.

[37] J. V. Dura, Development of new technologies for the flexible and eco-efficient production of customized healthy clothing, footwear and orthotics for consumers with high individualised needs, Fashon-able, Final Report, 2014, Information on http://www.fashionable-project.eu/

[38] E. Rosenblad-Wallin, User-oriented product development applied to functional clothing design, Appl. Ergon., vol. 16, no. 4, pp. 279–287, 1985. https://doi.org/10.1016/0003-6870(85)90092-4

[39] R. Ross, Home Economics- Fashion and Textile Technology Consumer Studies. Learning and Teaching Scotland, 2005.

[40] N. Shaari and N. Suleiman, Assistive Clothing for Disable People based on Kansei Approach Using Indigenous Clothing Construction, pp. 1–7, 2014.

[41] H. Meinander and M. Varheenmaa, Clothing and textiles for disabled and elderly people, VTT Tied. - Valt. Tek. Tutkimusk., no. 2143, pp. 3–57, 2002.

[42] S. Bragança, I. Castellucci, S. Gill, P. Matthias, M. Carvalho, and P. Arezes, Insights on the apparel needs and limitations for athletes with disabilities: The design of wheelchair rugby sports-wear, Appl. Ergon., vol. 67, pp. 9–25, 2018. https://doi.org/10.1016/j.apergo.2017.09.005

[43] Centres for Disease Control and Prevention, National Health and Nutrition Examination Survey (NHANES) III: Body Measurements (Anthropometry), vol. 20850, no. October. 1988.

[44] A.G.I.R.- S.I.T., Proiectarea constructivă a produselor de îmbrăcăminte, in Manualul Inginerului Textilist– Confectii Textile, Bucuresti: Ed. Agir, 2003, pp. 785–2091.

[45] Human Solutions and Hohensteiner Institute, Size Germany, Body measurements – Measurement Methodology in Garment Construction.

[46] I. Dabolina, Anthropometrical Measurements for Three-Dimensional Clothing Design, Proc. 1st Int. Conf. 3D Body Scanning Technol., no. October, p. 404, 2010. https://doi.org/10.15221/10.404

[47] N. D'Apuzzo, 3D body scanning technology for fashion and apparel industry, vol. 6491, p. 64910O, 2007. https://doi.org/10.1117/12.703785

[48] C. Niculescu, A. Salisteanu, E. Visileanu, E. Filipescu, and M. Avadanei, Baza de date 3d antropometrice, management şi aplicaţii, Buletinul Agir, Retrieved from: https://www.buletinulagir.agir.ro/articol.php?id=904

[49] L. W. W. Laing, An introduction to 3D Scanning, Comput. Des., vol. 26, no. 2, pp. 157–158, 1994. https://doi.org/10.1016/0010-4485(94)90038-8

[50] B. Breuckmann, 25 Years of High Definition 3D Scanning: History, State of the Art, Outlook, EVA London 2014 Proc. EVA London 2014 Electron. Vis. Arts, pp. 262–266, 1826. https://doi.org/10.14236/ewic/EVA2014.62

[51] J. Straub and S. Kerlin, Development of a Large, Low-Cost, Instant 3D Scanner, Technologies, vol. 2, no. 2, pp. 76–95, 2014. https://doi.org/10.3390/technologies2020076

[52] H. M. Daanen and G. J. van de Water, Whole body scanners, Displays, vol. 19, no. 3, pp. 111–120, 1998. https://doi.org/10.1016/S0141-9382(98)00034-1

[53] EinScan-SE. Information on: http://en.shining3d.com/.

[54] Artec Eva. Information on: https://www.artec3d.com/de/portable-3d-scanners/artec-eva.

[55] Artec Space Spider. Information on: https://www.artec3d.com/portable-3d-scanners/artec-spider.

[56] SYMCAD III. Information on: http://www.symcad.com/EXPORT/index_eng.htm.

[57] Aniwaa, 3D scanner categories and 3D scanner types. Information on: https://www.aniwaa.com/3d-scanners-categories/.

[58] P. R. Apeagyei, Application of 3D body scanning technology to human measurement for clothing Fit, Int. J. Digit. Content Technol. its Appl., vol. 4, no. 7, pp. 58–68, 2010. https://doi.org/10.4156/jdcta.vol4.issue7.6

[59] Artec, Artec support center. Information on: https://artecgroup.zendesk.com/hc/en-us.

[60] Artec 3D Scanner: Eva hilft bei der Erstellung einer 3D-Prothese eines zweiköpfigen Wadenmuskels. Information on: http://www.3d-model.ch/artec-3d-scanner-eva-hilft-bei-der-erstellung-einer-3d-prothese-eines-zweikoepfigen-wadenmuskels/.

[61]   3D Creative, Unique orthopedic scanning. Information on:
        http://3dcreative.lt/news/unique-orthopedic-scanning/?lang=en.

[62]   H. Löffler-Wirth et al., Novel anthropometry based on 3D-bodyscans applied to a
        large population based cohort, PLoS One, vol. 11, no. 7, pp. 1–20, 2016.
        https://doi.org/10.1371/journal.pone.0159887

[63]   Vialux, 360° Scan. Information on: https://www.vialux.de/en/360-scan.html.

[64]   H. A. M. Daanen and F. B. Ter Haar, 3D whole body scanners revisited, Displays,
        vol. 34, no. 4, pp. 270–275, 2013. https://doi.org/10.1016/j.displa.2013.08.011

[65]   3D Body scanner, Made-to-measure. Information on:
        http://www.bodyscan.human.cornell.edu/scene0f0a.html.

[66]   Human solutions, 3D Body Scanners. Information on: http://www.human-
        solutions.com/fashion/front_content.php?idcat=140&lang=7.

[67]   E. Jarosz, Determination of the workspace of wheelchair users, Int. J. Ind. Ergon.,
        vol. 17, no. 2, pp. 123–133, 1996. https://doi.org/10.1016/0169-8141(95)00044-5

[68]   B. Das and J. W. Kozey, Structural anthropometric measurements for wheelchair
        mobile adults, Appl. Ergon., vol. 30, no. 5, pp. 385–390, 1999.
        https://doi.org/10.1016/S0003-6870(99)00010-1

[69]   V. Sharma, Anthropometry of Indian manual wheelchair users: a validation study
        of Indian accessibility standards.Retrieved from:
        http://www.accessability.co.in/access/.

[70]   K. Lucero-Duarte, E. De La Vega-Bustillos, F. López-Millán, and S. Soto-Félix,
        Anthropometric data of adult wheelchair users for Mexican population, Work, vol.
        41, no. SUPPL.1, pp. 5408–5410, 2012. https://doi.org/10.3233/WOR-2012-0835-
        5408

[71]   V. Paquet and D. Feathers, An anthropometric study of manual and powered
        wheelchair users, Int. J. Ind. Ergon., vol. 33, no. 3, pp. 191–204, 2004.
        https://doi.org/10.1016/j.ergon.2003.10.003

[72]   K. Waugh and B. Crane, A clinical application guide to standardized wheelchair
        seating measures of the body and seating support surface, vol. 1. Colorado, USA:
        The Regents of the University of Colorado, 2013.

[73]   P. R. Bussell and L. L. F. Michaud, Seated anthropometry: The problems involved
        in a large scale survey of disabled and elderly people, Ergonomics, vol. 24, no. 11,
        pp. 831–845, 1981. https://doi.org/10.1080/00140138108924904

[74] D. A. Hobson and J. F. M. Molenbroek, Anthropometry and design for the disabled: Experiences with seating design for the cerebral palsy population, Appl. Ergon., vol. 21, no. 1, pp. 43–54, 1990. https://doi.org/10.1016/0003-6870(90)90073-7

[75] A. Hossein, D. Talab, A. B. Nezhad, and N. A. Darvish, Comparison of Anthropometric Dimensions in Healthy and Disabled Individuals, vol. 9, no. 3, 2017.

[76] B. Bradtmiller and J. Annis, Anthropometry for Persons With Disabilities : Needs for the Twenty-First Century, Washington, D.C., 1997

[77] B. Bradtmiller, Anthropometry of users of wheeled mobility aids: a critical review of recent work, Yellow Springs, OH, 2003

[78] N. Hernández, Tailoring the unique figure, Göteborg, 2000.

[79] E. Nowak, The role of anthropometry in design of work and life environments of the disabled population, Int. J. Ind. Ergon., vol. 17, no. 2, pp. 113–121, 1996. https://doi.org/10.1016/0169-8141(95)00043-7

[80] B. Brogin, D. C. Weiss, S. Marchi, M. Lucia, R. Okimoto, and S. T. De Oliveira, Advances in Ergonomics in Design, vol. 588, 2018.

[81] H. O. Barros and M. M. Soares, Anthropometric analysis of wheelchair users: Methodological factors which influence interpopulational comparison, Work, vol. 41, no. SUPPL.1, pp. 4091–4097, 2012. https://doi.org/10.3233/WOR-2012-0702-4091

[82] J. Park, Y. Choi, B. Lee, K. Jung, S. Sah, and H. You, A Classification of Sitting Strategies based on Driving Posture Analysis, Journal of the Ergonomics Society of Korea, vol. 33, no. 2, pp. 87–96, Apr. 2014. https://doi.org/10.5143/jesk.2014.33.2.87

[83] S. H. Kim, J. K. Pyun, and H. Y. Choi, Digital human body model for seat comfort simulation, Int. J. Automot. Technol., vol. 11, no. 2, pp. 239–244, 2010. https://doi.org/10.1007/s12239-010-0030-4

[84] 3D Systems. Information on: https://de.3dsystems.com/.

[85] InnovMetric Software Inc. Information on: https://www.innovmetric.com/en.

[86] S. Choi and S. P. Ashdown, 3D body scan analysis of dimensional change in lower body measurements for active body positions, Text. Res. J., vol. 81, no. 1, pp. 81–93, 2011. https://doi.org/10.1177/0040517510377822

[87] M. Najib, M. Lazim, and H. Lamsali, Body measurement using 3d handheld scanner, vol. 7, no. 1, pp. 179–187, 2018. https://doi.org/10.15282/mohe.v7i1.213

[88] Aniwaa, 3D body scanning, full body scanning and human body 3D scanners. Information on: https://www.aniwaa.com/3d-body-scanning/.

[89] A. Rudolf, A. Cupar, T. Kozar, and Z. Stjepanović, Study regarding the virtual prototyping of garments for paraplegics, Fibers Polym., vol. 16, no. 5, pp. 1177–1192, 2015. https://doi.org/10.1007/s12221-015-1177-4

[90] T. Kozar, A. Rudolf, S. Jevšnik, A. Cupar, and Z. Stjepanovič, Adapting human body model posture for the purpose of garment virtual prototyping, in Textile Science and Economy V, 5th International Scientific-Professional Conference, 2013, pp. 24–30.

[91] A. Rudolf et al., New technologies in the development of ergonomic garments for wheelchair users in a virtual environment In, Ind. Textila, vol. 68, no. 2, pp. 83–94, 2017. https://doi.org/10.35530/IT.068.02.1371

[92] GOM Precise Industrial 3D Metrology. Information on: https://www.gom.com/index.html.

[93] A. Cupar and A. Rudolf, Designing an Adaptive 3D Body Model Suitable for People with Limited Body Abilities, J. Text. Sci. Eng., vol. 4, no. 5, 2014. https://doi.org/10.4172/2165-8064.1000165

[94] S. Jevsnik, 3D Virtual Prototyping of Garments: Approaches, Developments and Challenges, J. Fiber Bioeng. Informatics, vol. 10, no. 1, pp. 51–63, 2017. https://doi.org/10.3993/jfbim00253

[95] A. Rudolf, S. Bogović, B. R. Car, A. Cupar, Z. Stjepanovič, and S. Jevšnik, Textile Forms' Computer Simulation Techniques, in Computer and Information Science, 2017, pp. 67–93. https://doi.org/10.5772/67738

[96] H. Koehler, M. Pruzinec, T. Feldmann, and A. Woerner, Automatic Human Model Parametrization From 3D Marker Data For Motion Recognition, WSCG'2008 - 16th Int. Conf. Cent. Eur. Comput. Graph. Vis. Comput. Vis., pp. 211–216, 2008.

[97] C. Meixner, Methodenentwicklung zur automatisierten Generierung virtuelle Bekleidungskonstruktion Danksagung, Fakultät Maschinenwesen der Technischen Universität Dresden, 2016.

[98] C. Meixner and S. Krzywinski, Development of a method for an automated generation of anatomy-based, kinematic human models as a tool for virtual

clothing construction, Comput. Ind., vol. 98, pp. 197–207, 2018.
https://doi.org/10.1016/j.compind.2018.03.003

[99]   L. Moccozet and F. Dellas, Animatable human body model reconstruction from 3d
       scan data using templates, 2004, Retrieved from:
       http://www.cs.uu.nl/centers/give/multimedia/publications/pdf/captech04.pdf

[100] P. Siebert and X. Ju, Animation Réaliste D ' Êtres Virtuels À Partir De Données
      Scannées Realistic Human Animation Using Scanned Data.

[101] N. M. Thalmann, F. Cordier, H. Seo, and G. Papagianakis, Modeling of Bodies and
      Clothes for Virtual Environments, Proc. 2004 Int. Conf. Cyberworlds, pp. 201–
      208, 2004.

[102] C. Meixner and S. Krzywinski, Development of a method for an automated
      generation of anatomy-based, kinematic human models as a tool for virtual
      clothing construction, Comput. Ind., vol. 98, pp. 197–207, 2018.
      https://doi.org/10.1016/j.compind.2018.03.003

[103] C. Meixner and S. Krzywinski, Automated Generation of Human Models from
      Scan Data in Anatomically Correct Postures for Rapid Development of Close-
      Fitting, Functional Garments, 2011, no. October, pp. 25–26.
      https://doi.org/10.15221/11.164

[104] C. Xu, J. He, X. Zhang, C. Yao, and P. H. Tseng, Geometrical kinematic modeling
      on human motion using method of multi-sensor fusion, Inf. Fusion, vol. 41, pp.
      243–254, 2018. https://doi.org/10.1016/j.inffus.2017.09.014

[105] K. Yamane and Y. Nakamura, Robot Kinematics and Dynamics for Modeling the
      Human Body, Robot. Res., vol. 66, pp. 49–60, 2010. https://doi.org/10.1007/978-3-
      642-14743-2_5

[106] Autodesk. Information on: https://www.autodesk.eu/.

[107] A. Leipner and S. Krzywinski, 3D Product Development Based on Kinematic
      Human Models, Proc. 4th Int. Conf. 3D Body Scanning Technol. Long Beach CA,
      USA, 19-20 Novemb. 2013, no. November, pp. 310–316, 2013.
      https://doi.org/10.15221/13.310

[108] J. Pan, L. Chen, Y. Yang, and H. Qin, Automatic skinning and weight retargeting
      of articulated characters using extended position-based dynamics, Vis. Comput.,
      pp. 1–13, 2017. https://doi.org/10.1007/s00371-017-1413-6

[109] L. Zhu, X. Hu, and L. Kavan, Adaptable Anatomical Models for Realistic Bone

Motion Reconstruction, Comput. Graph. Forum, vol. 34, no. 2, pp. 459–471, 2015. https://doi.org/10.1111/cgf.12575

[110] K. He, Rapid 3D Human Body Modeling and Skinning Animation Based on Single Kinect, J. Fiber Bioeng. Informatics, vol. 8, no. 3, pp. 413–421, 2015. https://doi.org/10.3993/jfbim00134

[111] A. Paquette, Introduction BT - An Introduction to Computer Graphics for Artists. 2013. https://doi.org/10.1007/978-1-4471-5100-5

[112] B. Allen, B. Curless, and Z. Popović, Articulated body deformation from range scan data, ACM Trans. Graph., vol. 21, no. 3, pp. 612–619, 2002. https://doi.org/10.1145/566654.566626

[113] D. Anguelov, P. Srinivasan, D. Koller, S. Thrun, J. Rodgers, and J. Davis, SCAPE: Shape Completion and Animation of People, *SIGGRAPH '05 ACM SIGGRAPH 2005 P*ap., pp. 408–416, 2005. https://doi.org/10.1145/1186822.1073207

[114] D. Hirshberg, M. Loper, E. Rachlin, and M.J. Black, in proceedings Hirshberg: ECCV:2012, Coregistration: Simultaneous alignment and modeling of articulated {3D} shape, booktitle: European Conf. on Computer Vision (ECCV), pages 242--255, series LNCS 7577, Part IV, editors A. Fitzgibbon et al. (Eds.), publisher Springer-Verlag, oct., 2012, 10. https://doi.org/10.1007/978-3-642-33783-3_18

[115] Y. Chen, Z. Liu, and Z. Zhang, Tensor-based human body modeling, Proc. IEEE Comput. Soc. Conf. Comput. Vis. Pattern Recognit., pp. 105–112, 2013. https://doi.org/10.1109/CVPR.2013.21

[116] G. Pons-Moll, J. Romero, N. Mahmood, and M. J. Black, Dyna: A Model of Dynamic Human Shape in Motion, ACM Trans. Graph., vol. 34, no. 4, pp. 120:1-120:14, 2015. https://doi.org/10.1145/2766993

[117] The Dyna model, https://www.youtube.com/watch?v=mWthea2K8-Q.

[118] C. Forza, P. Romano, and A. Vinelli, Information Technology for Managing the Textile Apparel Chain: Current Use, Shortcomings and Development Directions, Int. J. Logist. Res. Appl., vol. 3, no. 3, pp. 227–243, 2000. https://doi.org/10.1080/713682770

[119] A. S. M. Sayem, R. Kennon, and N. Clarke, 3D CAD systems for the clothing industry, Int. J. Fash. Des. Technol. Educ., vol. 3, no. 2, pp. 45–53, 2010. https://doi.org/10.1080/17543261003689888

[120] Y. J. Liu, D. L. Zhang, and M. M. F. Yuen, A survey on CAD methods in 3D

garment design, Comput. Ind., vol. 61, no. 6, pp. 576–593, 2010.
https://doi.org/10.1016/j.compind.2010.03.007

[121] Y. Meng, P. Y. Mok, and X. Jin, Computer aided clothing pattern design with 3D editing and pattern alteration, CAD Comput. Aided Des., vol. 44, no. 8, pp. 721–734, 2012. https://doi.org/10.1016/j.cad.2012.03.006

[122] S. Jevsnik, T. Pilar, Z. Stjepanovic, and a. Rudolf, Virtual Prototyping of Garments and Their Fit to the Body, Daaam Int. Sci. B. 2012, pp. 601–618, 2012. https://doi.org/10.2507/daaam.scibook.2012.50

[123] M. Mahnic Naglic, S. Petrak, and Z. Stjepanovič, Analysis of Tight Fit Clothing 3D Construction Based on Parametric and Scanned Body Models, Proc. 7th Int. Conf. 3D Body Scanning Technol. Lugano, Switzerland, 30 Nov.-1 Dec. 2016, no. December, pp. 302–313, 2016. https://doi.org/10.15221/16.302

[124] D. Zhang, Y. Liu, J. Wang, and J. Li, An integrated method of 3D garment design, J. Text. Inst., vol. 5000, pp. 1–11, 2018.

[125] R. Raffaeli and M. Germani, Knowledge-based approach to flexible part design, J. Eng. Des., vol. 21, no. 1, pp. 7–29, 2010. https://doi.org/10.1080/09544820802086996

[126] Gerber. Information on: http://www.gerbertechnology.com/.

[127] Gemini CAd Systems. Information on: https://www.geminicad.com/.

[128] Lectra. Information on: https://www.lectra.com/en.

[129] E. Papahristou, The effective integration of 3D virtual prototype in the product development process of the textile/clothing industry, Technical University of Crete School, 2016.

[130] D. Zhang, Y. Liu, J. Wang, and J. Li, An integrated method of 3D garment design, J. Text. Inst., vol. 5000, pp. 1–11, 2018.

[131] P. Volino, F. Cordier, and N. Magnenat-Thalmann, From early virtual garment simulation to interactive fashion design, *CAD* Comput. Aided Des., vol. 37, no. 6, pp. 593–608, 2005. https://doi.org/10.1016/j.cad.2004.09.003

[132] Z. G. Luo and M. M. F. Yuen, Reactive 2D/3D garment pattern design modification, CAD Comput. Aided Des., vol. 37, no. 6, pp. 623–630, 2005. https://doi.org/10.1016/j.cad.2004.09.005

[133] H. Rödel, A. Schenk, C. Herzberg, and S. Krzywinski, Links Between Design,

Pattern Development and Fabric Behavior for Clothing and Technical Textiles, Tech. Text., vol. 1, no. 4, pp. 1–8, 2001. https://doi.org/10.1108/EUM0000000005782

[134] S. Krzywinski, A. Schenk, J. Siegmund, and H. Rödel, Virtual Product Development for Garments and Technical Textiles, in 6th World Textile Conference AUTEX, 2006.

[135] J. Siegmund, S. Krzywinski, E. Kirchdörfer, and A. Mahr-Erhardt, Development of parametric virtual dummies. Scalable forms for the torso, Text. Netw., no. 6, pp. 17–21, 2008.

[136] E. Chaw Hlaing, S. Krzywinski, and H. Roedel, Garment prototyping based on scalable virtual female bodies, Int. J. Cloth. Sci. Technol., vol. 25, no. 3, pp. 184–197, 2013. https://doi.org/10.1108/09556221311300200

[137] S. Krzywinski and J. Siegmund, 3D Product Development for Loose-Fitting Garments Based on Parametric Human Models, IOP Conf. Ser. Mater. Sci. Eng., vol. 254, no. 15, 2017. https://doi.org/10.1088/1757-899X/254/15/152006

[138] E. Chaw Hlaing, Development of Reproducible Methods of Construction for Loose-fitting Garments on the Basis of 3D Virtual Female Models, TU Dresden, 2013.

[139] S. Krzywinski, Verbidung von Design und Konstruktion in der textilen Konfektion unter Anwendung von CAE, TU Dresden, 2005.

[140] J. Siegmund, Erarbeitung virtueller Menschmodelle als Konstruktionswerkzeug zur 3D Produktentwicklung in der Bekleidungsindustrie, TU Dresden, 2013.

[141] S. Krzywinski, J. Siegmund, and H. Rödel, Design of body-fitted knitwear garments, Text. Netw., no. 3, pp. 18–20, 2005.

[142] S. Krzywinski, A. Schenk, E. Haase, S. Papst, and B. Thomaszewski, Simulation and virtual fit control. Closer to reality., Text. Netw., no. 10, pp. 30–36, 2008.

[143] E. Chaw Hlaing, S. Krzywinski, and H. Roedel, Garment prototyping based on scalable virtual female bodies, Int. J. Cloth. Sci. Technol., vol. 25, no. 3, pp. 184–197, 2013. https://doi.org/10.1108/09556221311300200

[144] S. Olaru, E. Filipescu, M. Avadanei, A. Mocenco, G. Popescu, and A. Săliştean, Applied 3D virtual try-on for bodies with atypical characteristics, Procedia Eng., vol. 100, no. January, pp. 672–681, 2015. https://doi.org/10.1016/j.proeng.2015.01.419

[145] C. Paper, Z. Stjepanovi, B. Luka, T. View, and B. C. View, Applying CASP method for construction of adapted garments for people with scoliosis, no. November, 2015.

[146] Y. Hong, P. Bruniaux, X. Zeng, K. Liu, A. Curteza, and Y. Chen, Visual-simulation-based Personalized Garment Block Design Method for Physically Disabled People with Scoliosis (PDPS), Autex Res. *J.*, vol. 0, no. 0, pp. 1–11, 2017. https://doi.org/10.1515/aut-2017-0001

[147] N. Pruthi, C. Pruthi, and P. Sutharamn, Protective Clothing for Paraplegic Men, J. Hum. Ecol., vol. 20, no. 2, pp. 103–108, 2006. https://doi.org/10.1080/09709274.2006.11905911

[148] Y. Wang, D. Wu, M. Zhao, and J. Li, Evaluation on an ergonomic design of functional clothing forwheelchair users, Appl. Ergon., vol. 45, no. 3, pp. 550–555, 2014. https://doi.org/10.1016/j.apergo.2013.07.010

[149] W.-M. Chang, Y.-X. Zhao, R.-P. Guo, Q. Wang, and X.-D. Gu, Design and Study of Clothing Structure for People with Limb Disabilities, J. Fiber Bioeng. Informatics, vol. 2, no. 1, pp. 62–67, 2009. https://doi.org/10.3993/jfbi06200910

[150] Artec ™ MH and MHT 3D Scanners. Information on: http://www.microgeo.it/public/userfiles/spagnolo/download-es/Artec-MH-MHT-3D.pdf.

[151] Vialux, zSnapper ® cart specifications. Information on: https://www.vialux.de/en/360-scan.html.

[152] GeomagicStudio. Information on: https://de.3dsystems.com/.

[153] 3dsMax. Information on: https://www.autodesk.com/products/3ds-max/overview.

[154] B. W. Pastorius, Triangulation sensors, An Overview, Journal, pp. 1–12, 1971.

[155] W. P. Kennedy, The Basics of Triangulation Sensors, Sensors, Peterborough, NH, no. May 1998, pp. 1–8, 2005.

[156] M. F. M. Costa, Optical triangulation-based microtopographic inspection of surfaces, Sensors, vol. 12, no. 4, pp. 4399–4420, 2012. https://doi.org/10.3390/s120404399

[157] P. Rodríguez Gonzálvez, Á. L. Muñoz Nieto, S. Zancajo Blázquez, and D. González Aguilera, Geomatics and Forensic: Progress and Challenges, in Forensic Analysis - From Death to Justice, INTECH, 2016. https://doi.org/10.5772/63155

[158] S. Hwang, Three Dimensional Body Scanning Systems With Potential for Use in the Apparel Industry, North Carolina State University, 2001.

[159] I. Álvarez, J. M. Enguita, M. Frade, J. Marina, and G. Ojea, On-line metrology with conoscopic holography: Beyond triangulation, Sensors, vol. 9, no. 9, pp. 7021–7037, 2009. https://doi.org/10.3390/s90907021

[160] Artec 3D, Artec Studio Userguide.

[161] J. Wilm, H. Aanæs, R. Larsen, and R. R. Paulsen, Real Time Structured Light and Applications, Kongens Lyngby - Technical University of Denmark, 2016.

[162] S. Foster and D. Halbstein, Integrating 3D Modeling, Photogrammetry and Design, 2014. https://doi.org/10.1007/978-1-4471-6329-9

[163] J. Awange and E. Sholarin, Photogrammetry, in Environmental Project Management, no. JANUARY, Switzerland: Springer International Publishing, 2015, pp. 367–371. https://doi.org/10.1007/978-3-319-27651-9_17

[164] M. a. Brunsman, H. A. M. Daanen, and K. M. Robinette, Optimal postures and positioning for human body scanning, Proceedings. Int. Conf. Recent Adv. 3-D Digit. Imaging Model. (Cat. No.97TB100134), pp. 266–273, 1997.

[165] K. Chang, A Review on Shape Engineering and Design Parameterization in Reverse Engineering, Reverse Eng. – Recent Adv. Appl., pp. 161–186, 11AD.

[166] 3D Systems, Geomagic Qualify. Inforation on: https://de.3dsystems.com/press-releases/geomagic/qualify-enables-graphical-comparisons-between-cad-maste.

[167] Y. Meng, P. Y. Mok, and X. Jin, Interactive virtual try-on clothing design systems, CAD Comput. Aided Des., vol. 42, no. 4, pp. 310–321, 2010. https://doi.org/10.1016/j.cad.2009.12.004

[168] Y. Hong, P. Bruniaux, X. Zeng, K. Liu, A. Curteza, and Y. Chen, Visual-simulation-based Personalized Garment Block Design Method for Physically Disabled People with Scoliosis (PDPS), Autex Res. J., vol. 18, no. 1, pp. 1–11, 2017. https://doi.org/10.1515/aut-2017-0001

[169] Y. A. N. Hong, P. Bruniaux, J. Zhang, K. Liu, M. I. N. Dong, and Y. A. N. Chen, Application of 3D-TO-2D garment design for atypical morphology : a design case for physically disabled people with scoliosis, vol. 69, no. 1, pp. 59–64, 2018. https://doi.org/10.35530/IT.069.01.1377

[170] Lectra, Information on: https://www.lectra.com/en.

[171] K. Liu, E. Kamalha, J. Wang, and T. K. Agrawal, Optimization design of cycling clothes' patterns based on digital clothing pressures, Fibers Polym., vol. 17, no. 9, pp. 1522–1529, 2016. https://doi.org/10.1007/s12221-016-6402-2

[172] S. Kim, Y. Jeong, Y. Lee, and K. Hong, 3D pattern development of tight-fitting dress for an asymmetrical female manikin, Fibers Polym., vol. 11, no. 1, pp. 142–146, 2010. https://doi.org/10.1007/s12221-010-0142-5

[173] Y. Jeong, K. Hong, and S. J. Kim, 3D pattern construction and its application to tight-fitting garments for comfortable pressure sensation, Fibers Polym., vol. 7, no. 2, pp. 195–202, 2006. https://doi.org/10.1007/BF02908267

[174] J. P. da Silva, A. L. A. Júnior, and G. A. Giraldi, A Review of Dynamic NURBS Approach, 2013. https://doi.org/10.1016/j.jksuci.2014.12.010

[175] D. Saini, S. Kumar, and T. R. Gulati, NURBS-based geometric inverse reconstruction of free-form shapes, J. King Saud Univ. - Comput. Inf. Sci., vol. 29, no. 1, pp. 116–133, 2017.

[176] Clo 3D, Information on: https://www.clo3d.com/.

[177] M. Scott, Pattern cutting for clothing using CAD - How to use Lectra Modaris, 2012. https://doi.org/10.1533/9780857097095

[178] Optitex, Information on: https://optitex.com/.

[179] V-Stitcher, Information on: https://browzwear.com/products/v-stitcher/.

[180] K. Liu, X. Zeng, P. Bruniaux, J. Wang, E. Kamalha, and X. Tao, Fit evaluation of virtual garment try-on by learning from digital pressure data, Knowledge-Based Syst., vol. 133, pp. 174–182, 2017. https://doi.org/10.1016/j.knosys.2017.07.007

[181] A. Porterfield and T. A. M. Lamar, Examining the effectiveness of virtual fitting with 3D garment simulation, Int. J. Fash. Des. Technol. Educ., vol. 10, no. 3, pp. 320–330, 2017. https://doi.org/10.1080/17543266.2016.1250290

[182] F. Z. Shuixian Hu, Ruomei Wang, An efficient multi-layer garment virtual fitting algorithm based on the geometric method, Int. J. Cloth. Sci. Technol., vol. 29, no. 1, pp. 25–38, 2017. https://doi.org/10.1108/IJCST-06-2015-0068

[183] H. Jedda, A. Ghith, and F. Sakli, Prediction of fabric drape using the FAST system, J. Text. Inst., vol. 98, no. 3, pp. 219–225, 2007. https://doi.org/10.1080/00405000701463920

[184] A. N. Ali, Using Siro FAST System to Measure Handle Properties of Outerwear

Woven from monofilaments Polyester yarn, Life Sci. J., vol. 12, no. 6, pp. 129–134, 2015.

[185] Fast system fabric assurance by simple testing, Retrieved from: http://nptel.ac.in/courses/116102029/56.

[186] H. Eberle, H. Hermeling, M. Hornberger, D. Menzer, and W. Ring, Clothing Technology, Second Edi. Haan-Gruiten: Europa Lehrmittel, 1999.

Digital Methods in Developing Textile Products for People with Locomotor Disabilities
Materials Research Foundations **110** (2021)

Materials Research Forum LLC
https://doi.org/10.21741/9781644901557

Abbreviations

BMI – Body Mass Index
C - Cervical
CAD – Computer Aided Design
CSIRO – Commonwealth Scientific and Industrial Research Organization
D – Depth
DLP – Digital Light Processing
DMD – Digital Micromirror Device
DXF – Drawing Exchange Format
EU – European union
FAST – Fabric Assurance by Simple Testing
H - Height
HSP – Hereditary Spastic Paraplegia
ICF – International Classification of Functioning
IGES – Initial Graphics Exchange Format
ICIDH – International Classification of Impairments, Disabilities and Handicaps
L – Lumbar
MEMS – Micro – Electromechanical System
MRI – Magnetic Resonance Imaging
NURB – Non-Uniform Rational B-Splines
OBJ – Object File
S – Sacral
SCI – Spinal Cord Injury
SPG – Spastic Paraplegia
STL – Standard Template Library
T – Thoracic
W – Width
WHO – World Health Organization
2D – Two-dimensional
3D – Three-dimensional

About the Authors

Bianca Aluculesei is a textile engineer, working in the clothing industry since 2013. In 2014 she decided to enter the research field and she began the doctoral studies willing to analyse what necessities the paraplegic persons have, regarding their clothing. In 2019, after a fruitful collaboration between The "Gheorghe Asachi" Technical University of Iasi, from Romania, and the Technical University of Dresden, from Germany, she finished her PhD dissertation under the supportive guidance of her supervisors. Her doctoral thesis, "Researches on the implementation of new digital methods for the development of textile products for people with locomotor disabilities." provides new methods for the digital development of textile products for wheelchair users.

Sybille Krzywinski studied at Technischen Universität Dresden, Germany, in the area of Textile Technology, and her habilitation theses was entitled "3D product development for garments and technical textiles by means of CAE". She is currently Professor dr. eng. habil. at Technical University of Dresden, Institute of Textile Machinery and High Performance Material Technology, Germany. Between 2016 and 2019 she was an acting director at the chair of "Ready-made Technology", and from 2019 she is the scientific director at the chair "Development and Assembly Technology for Textile Products". She taught in areas of Spinning, Weaving, Ready-made Technology, Design and Construction/CAD, 3D-CAD-Applications - Virtual product development, Digital process chains, Technical Textiles. She published more than 230 research articles in professional journals, and presented lectures at national and international conferences. Professor Sybille Krzywinski has coordinated, developed and managed numerous public funded research and industrial projects in the field of aircraft and vehicle construction, sports and medical textiles/garments, and conventional textiles/clothing. Since 2016 she is Equal Opportunity Commissioner at the Faculty of Mechanical Engineering. She is also reviewer for refereed textile journals.

Antonela Curteza graduated the Faculty of Textile and Garment Engineering from *the* "Gheorghe Asachi" Technical University of Iasi, Romania, and holds a Doctorate degree in Functions and Comfort of Clothing. She is Professor dr. eng. habil at the Knitting and Clothing Engineering Department within the Faculty of Industrial Design and Business Management. She has initiated and conducted many research and teaching activities focused mainly on: design, clothing comfort and functions, *s*mart and functional textile products, sustainable fashion. Professor Antonela Curteza has been a PhD supervisor and scientific coordinator of 15 PhDs in the field of Industrial Engineering. She is author and co-author of 12 books having garment design, functions and comfort as main subjects. She has authored more than 170 articles published by international and national journals, and international conference proceedings, from which 55 are ISI certified. She took part in 21 national and 9 international research projects; in 16 of them she held the position of coordinator or partner manager. The main current research work and interests are: clothing comfort and functions, sustainable development in fashion and design, functional textile products for people with special needs and protective clothing.

Manuela Avadanei is an Associate Professor at The "Gheorghe Asachi" Technical University of Iasi-Romania, Faculty of Industrial Design and Business Management, with a PhD degree in the field. She is an expert in the field of anthropometry applied in textile engineering, with an emphasis on obtaining, analysing and interpreting information regarding the shape of the human body, which is subsequently employed in the ergonomic pattern making process of clothing (fashion products, protective or medical equipment, sportswear, etc.). She has also focused on the pattern making process for complex-shaped and highly personalized apparel products, which involves 2D and 3D IT tools (specialized CAD systems for the clothing industry), based on the customer's needs and requirements, as part of a digitalized fashion and clothing industry. She continues her research and academic activity in pattern making by exploring new ideas or techniques for developing ergonomic garment products adapted to different demands, body shapes, and purposes (fashionable, sportive or protective). The author is also actively involved in research projects and in publishing articles in the aforementioned areas.